Management and Industrial Engi

Series editor

J. Paulo Davim, Aveiro, Portugal

More information about this series at http://www.springer.com/series/11690

J. Paulo Davim

Editor

Progress in Lean Manufacturing

 Springer

Editor
J. Paulo Davim
Department of Mechanical Engineering
University of Aveiro
Aveiro
Portugal

ISSN 2365-0532   ISSN 2365-0540   (electronic)
Management and Industrial Engineering
ISBN 978-3-319-89253-5   ISBN 978-3-319-73648-8   (eBook)
https://doi.org/10.1007/978-3-319-73648-8

Printed on acid-free paper

This Springer imprint is published by the registered company Springer International Publishing AG
part of Springer Nature
The registered company address is: Gewerbestrasse 11, 6330 Cham, Switzerland

Preface

Lean thinking was a concept developed by James P. Womack and Daniel T. Jones to capture the essence of Toyota Production System. Therefore, it is current report lean thinking as a lean way of thinking allows companies to '*specify value, line up value creating actions in the best sequence, conduct these activities without interruption whenever someone requests them, and perform them more and more effectively*'. This declaration leads to the five principles of lean thinking: *Value, Value Stream, Flow, Pull* and *Perfection*. The concept of lean thinking presents great importance in the context of modern manufacturing.

The purpose of this book is to present a collection of chapters exemplifying progress in lean manufacturing. The first chapter of the book provides leanness assessment tools and frameworks. The second chapter is dedicated to lean supply chain management (a systematic literature review of practices, barriers and contextual factors inherent to its implementation). The third chapter describes a literature review on lean manufacturing in small manufacturing companies. The fourth chapter contains information on application of structural equation modelling for analysis of lean concepts deployment in healthcare sector. Finally, the last chapter is dedicated to association between lean manufacturing teaching methods and students' learning preferences.

The current book can be used as a research book for final undergraduate engineering course or as a topic on management and industrial engineering at the postgraduate level. Also, this book can serve as a useful reference for academics, engineers, managers, researchers, professionals in management and industrial engineering and related subjects. The interest of scientific in this book is evident for many important centers of the research and universities as well as industry. Therefore, it is hoped this book will inspire and enthuse others to undertake research in management and industrial engineering.

The Editor acknowledges Springer for this opportunity and professional support. Finally, I would like to thank all the chapter authors for their availability for this editorial project.

Aveiro, Portugal   J. Paulo Davim
January 2018

Contents

About the Editor

J. Paulo Davim received the Ph.D. degree in Mechanical Engineering in 1997, the M.Sc. degree in Mechanical Engineering (materials and manufacturing processes) in 1991, the Mechanical Engineer degree (MEng-5 years) in 1986, from the University of Porto (FEUP), the Aggregate title (Full Habilitation) from the University of Coimbra in 2005 and the D.Sc. from London Metropolitan University in 2013. He is Eur Ing by FEANI-Brussels and Senior Chartered Engineer by the Portuguese Institution of Engineers with a MBA and Specialist title in Engineering and Industrial Management. Currently, he is Professor at the Department of Mechanical Engineering of the University of Aveiro, Portugal. He has more than 30 years of teaching and research experience in Manufacturing, Materials and Mechanical Engineering with special emphasis in Machining & Tribology. He has also interest in Management & Industrial Engineering and Higher Education for Sustainability & Engineering Education. He has guided large numbers of postdoc, Ph.D. and masters students. He has received several scientific awards. He has worked as evaluator of projects for international research agencies as well as examiner of Ph.D. thesis for many universities. He is the Editor in Chief of several international journals, Guest Editor of journals, books Editor, book Series Editor and Scientific Advisory for many international journals and conferences. Presently, he is an Editorial Board member of 25 international journals and acts as reviewer for more than 80 prestigious Web of Science journals. In addition, he has also published as editor (and co-editor) more than 100 books and as author (and co-author) more than 10 books, 70 book chapters and 400 articles in journals and conferences (more than 200 articles in journals indexed in Web of Science core collection/h-index 43+/5500+ citations and SCOPUS/h-index 52+/7500+ citations).

Leanness Assessment Tools and Frameworks

Omogbai Oleghe and Konstantinos Salonitis

Abstract This chapter presents the most recent developments with regards the assessment of leanness in manufacturing organizations. Leanness is the measure of the performance of lean manufacturing practices. It is tracked for improvement using assessment frameworks. This chapter reviews prevalent frameworks in order to organize the knowledge, extract the typical and potential uses, establish strengths and weaknesses and reveal ways of improving the extant frameworks. Prevailing frameworks are identified through a search of literature, together with those developed by lean consultants, as well as award-based frameworks. Two main classification schemes are used to organize and compare the frameworks namely the leanness indicators (input data type) used in the frameworks and the applications of the frameworks, representing the inputs and outputs respectively of the frameworks. The key findings of this work can be summarized into: First, most frameworks are generated using either a quantitative or qualitative set of leanness indicators; meanwhile there is a paucity of frameworks that use both types of indicators simultaneously to take advantage of their individual strengths and overcome their respective weaknesses. Second, the frameworks have been used mainly for current-as-is audits, whereas the assessment of proposed improvements is rarely considered. Third, majority of frameworks do not emphasize the interactions between lean practices and the trade- offs between their improvements.

Keywords Leanness · Leannes assessment · Lean maturity · Lean indicators · Performance measurement

O. Oleghe · K. Salonitis (✉)
Manufacturing Theme, Cranfield University, Cranfield MK43 0AL, UK
e-mail: k.salonitis@cranfield.ac.uk

© Springer International Publishing AG, part of Springer Nature 2018
J. P. Davim (ed.), *Progress in Lean Manufacturing*,
Management and Industrial Engineering,
https://doi.org/10.1007/978-3-319-73648-8_1

1 Introduction

Lean manufacturing (LM) takes its roots from manufacturing best-practices that were implemented in the Toyota Motor Corporation, such as Just in Time (JIT) management, Quality Management (QM), Total Productive Maintenance (TPM). LM has evolved over the years to include a variety of management values related with Employee Involvement, Supplier Management, Cross-functional Teams, Training, Customer Engagement and many others (Shah and Ward 2003, 2007), with a variety of tools such as Six Sigma, Statistical Process Control, Poka-Yoke, Jidoka.

Lean practices have been proven to improve manufacturing and organizational performance. The practices are intended to achieve multiple objectives for an organization, chiefly to improve customer responsiveness through continuous improvement and identification/elimination of all types of activities and processes that do not add to customer value. A collection of lean practices constitutes a lean system: the practices cannot be individually adopted on their own if an overall lean state is to be attained (Hallam 2003a; Rymaszewska 2014). There is in fact limited positive impact on performance when lean practices are introduced in isolation (Bonavia and Marin 2006).

Leanness is a concept that unifies the various practices of promoting lean (Bayou and De Korvin 2008). Leanness assessment is the measure of the adoption of lean manufacturing practices (Susilawati et al. 2013; Vimal and Vinodh 2013). Leanness assessment refers to the structured approach taken to assess leanness level as represented in Fig. 1.

The hierarchical structure shown in Fig. 1 represents the key components of the methodological steps, as well as the sequence and levels of assessment. In the configuration, the various lean practices and their measurement items are singled out and assessed using one or a set of tools and instruments, from which

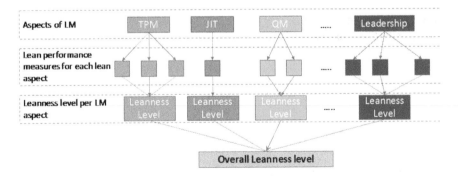

Fig. 1 Generic structured approach to leanness assessment

information about the lean state of the system can be generated. Lean Enterprise Self-Assessment Tool (LESAT) is a good example. Aspects of lean transformation are set apart in three sections under the LESAT framework namely Leadership, Processes and Infrastructure (LAI 2012). The indicators of leanness are the 54 measurement items under each of the three sections, for example, one measurement item under Leadership is the extent to which the organization integrates enterprise transformation into its strategic planning process. The analysis of the gap between the current and desired levels of performance for each measurement item is indicative of the leanness level for the measured practice (Perkins et al. 2010). The final tallies, ranges, patterns and averages provide information about the state of leanness of the enterprise (Perkins et al. 2010). Majority of methods can be described using this generic format also depicted in Fig. 1. The structured approach of leanness assessment helps to reduce chaos in terms of which lean practices to implement (Cil and Turkan 2013), where improvement efforts should be focused (Vinodh and Balaji 2011; Vimal and Vinodh 2013) and assist in the decision-making process (Chhabi et al. 2014).

The critique of leanness assessment methods has been undertaken in the past by various authors (Ray et al. 2006; Wan and Chen 2008; Mahfouz 2011; Anvari et al. 2013; Cil and Turkan 2013; Chhabi et al. 2014; Azadeh et al. 2015; Ali and Deif 2016). Most of these reviews have been limited in scope whereby only a handful of methods are appraised. If previous reviews have been done scantily, it implies that there is limited knowledge about what the expansive range of methods have accomplished. If there is limited knowledge about what leanness assessment methods can achieve, then lean practitioners and academics are not fully aware of what is available for them to use. Meanwhile, a number of tools and instruments have been developed for the leanness assessment. They have been applied singly and in a mixed manner. They have been validated to show the multiple benefits that can be derived from their use. Yet the knowledge is not organized. Meanwhile, it could be argued that the literature is representative of what is used in industry, since majority of the study methods were validated in real life cases. In addition, some methods coming from lean groups and consultants share similarities with what is available in the literature. For example, the Gemba Academy (Gemba 2010) have developed a lean enterprise self-assessment tool that is similar to methods used in various studies (Vinodh and Balaji 2011; Vimal and Vinodh 2013). The Strategos LAT developed by Quaterman Lee (Strategos 2010) has been used in multiple studies (Taj 2005; Ihezie and Hargrove 2009). The Association for Manufacturing Excellence (AME) LAT is based on the Iwao Kobayashi's 20 Keys to workplace improvement (AME 2016).

The current chapter aims to survey leanness assessment methods that prevail in literature. The survey will focus on the key aspects addressed by the methods. In addition, the survey will investigate the key instruments and data types that are used in the methods. The intention is to reveal common themes and trends as well as gaps, to provide directions where future advancements can be made in the methods.

2 Search Strategy to Generate List of Leanness Assessment Tools

Various academic databases and search engines such as Springerlink, Google Scholar, ABI/Inform Complete, EBSCO, Elsevier, Emerald Full Text, Science Direct, Scopus, and Taylor and Francis were consulted to extract relevant articles from which the methods were set apart. Search keywords included leanness, manufacturing leanness, leanness assessment and lean assessment tools. The relevant articles covered the period between 2000 and 2016, and did not limit the search to high impact journals only, in order to capture as many studies that are relevant as well as current. In fact, some studies with interesting results were found from journals with low to average impact factor. The adopted search strategy generated 64 relevant research publications from which a comprehensive list of leanness assessment methods was extracted.

The relevant articles were surveyed in three parts (Fig. 2). The first part was used to summarize the key uses of the methods. The second part was used to outline the tools/instruments that were used in the methods. The final part details the data types used in the assessment.

3 Key Uses and Outputs of the Leanness Assessment Methods

The uses of leanness assessment methods have been varied even though past researchers/users all make reference to assessing leanness. An uninformed and casual observer at first glance of the literature will be overwhelmed or confused with the variety of the studies, study outcomes and uses of the methods. In this section, the generic study outcomes (those concerned with assessing manufacturing leanness) are revealed to provide details of the typical uses of the various methods that have been advanced. Seven general or common uses have been found for the methods namely: quantification of leanness level, gap analysis, impact analysis, degree of adoption of lean practices, benchmarking, scenario analysis and dynamic analysis (Fig. 3).

Fig. 2 Leanness Assessment (LA) analysis structure

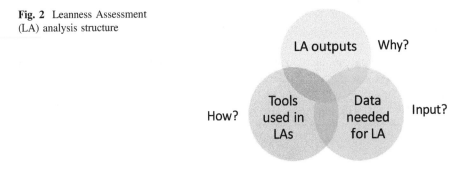

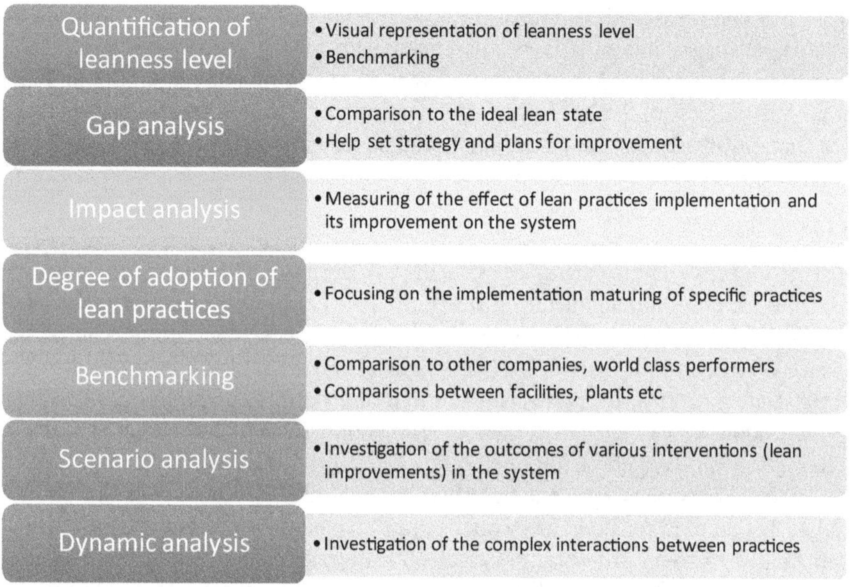

Quantification of leanness level	• Visual representation of leanness level • Benchmarking
Gap analysis	• Comparison to the ideal lean state • Help set strategy and plans for improvement
Impact analysis	• Measuring of the effect of lean practices implementation and its improvement on the system
Degree of adoption of lean practices	• Focusing on the implementation maturing of specific practices
Benchmarking	• Comparison to other companies, world class performers • Comparisons between facilities, plants etc
Scenario analysis	• Investigation of the outcomes of various interventions (lean improvements) in the system
Dynamic analysis	• Investigation of the complex interactions between practices

Fig. 3 Why lean assessment methods and tools are used

- **Quantification of leanness level**

Leanness level has been measured, quantified and represented by numbers. The quantification of leanness has been achieved using scoring methods, made possible through multi-level grading schemes (Taj 2005; Shetty et al. 2010). Leanness quantification has also been realized using the lean index (Ray et al. 2006; Wan 2006; Deif 2012; Vinodh and Vimal 2012; Berlec et al. 2014; Pakdil and Leonard 2014; Oleghe and Salonitis 2016a, b). The lean index can be defined as the weighted summation of the lean metrics that define performance of various variables representing lean practices within a system (Oleghe and Salonitis 2015). It is a single indicative score for overall lean performance (Searcy 2009).

Multiple benefits are derived from leanness quantification. The quantified leanness can be tracked using line graphs, statistical process control charts and radar charts (Taj 2005; Ray et al. 2006; Pakdil and Leonard 2014; Wong et al. 2014; Oleghe and Salonitis 2016a, b). The lean index can be used to drive organization-wide behaviour towards improving the metric (Searcy 2009; Wong et al. 2014), in a collaborative manner (Vimal and Vinodh 2012; Wong et al. 2014). Leanness quantification also enables objective benchmarking (Srinivasaraghavan and Allada 2006; Ray et al. 2006; Bayou and De Korvin 2008).

There are some limitations associated with leanness quantification. If a single metric is used to quantify leanness level, organizations may be tempted to use it exclusively (Wong et al. 2014). Rather, the metric should be decomposed for better understanding of its components i.e. the individual measures of lean (Searcy 2009).

In fact, a one-off quantification of the leanness level is insufficient to indicate the state of leanness, unless it is benchmarked with best practice (Ray et al. 2006; Srinivasaraghavan and Allada 2006) or tracked over time (Ray et al. 2006; Wong et al. 2014). Most of the leanness assessment frameworks stop at the leanness level per lean aspect step, as shown in Fig. 1, and do not proceed in the calculation of an aggregated overall leanness level.

- **Gap analysis**

A key objective of leanness assessment has been to investigate, evaluate and measure the current manufacturing situation vis-à-vis an ideal lean state (Ray et al. 2006). Various methods have been advanced to measure the gap in lean perfor-mance. The Lean Enterprise Self-Assessment Tool (LESAT) (Nightingale and Mize 2002; Hallam 2003a, b) uses a pair (current and desired) of performance ratings. Organizations score themselves on both ratings on a scale of 1–5 along 54 practices relating to lean. By subtracting the current score from the desired score, a gap can be calculated for each lean practice (Perkins et al. 2010). Large gaps are an indi-cation that there is considerable scope for improvement (Perkins et al. 2010). Small gaps imply a variety of issues; the assessed lean practice may not be so relevant in the company or the organization is yet to realize the importance of the practice (Perkins et al. 2010). The gap analysis with the LESAT can also be used to understand the level of agreement and cohesion amongst organizational members who filled out the responses in the LESAT (Perkins et al. 2010). Srinivasaraghavan and Allada (2006) used the Mahalanobis Taguchi Gram Schmidt System (MTGS) orthogonalization process to establish the degree of leanness abnormality or gap between an advanced lean group of companies and an average/below lean group. The method revealed specific improvements that can be initiated in a company in the average/below lean group. Wan et al. (2007) and Wan and Chen (2008) proposed a methodology based on Data Envelopment Analysis to determine the gap between current lean state and an ideally lean state, by using a virtual ideally lean Decision-Making Unit as the benchmark. The gap was used as a measure of lean performance for the system. Taj (2005) used a spreadsheet-based lean assessment tool developed by Quaterman Lee (Strategos LAT) to evaluate LM performance amongst Chinese hi-tech manufacturers, along nine key areas of manufacturing. The gap analysis was outputted to a lean profile chart (radar chart) to establish the current state of lean in the industry, their target, and the gap between the actual and the target. The results provided background information for further investigation of the industry as to why some companies performed better than others. Anand and Kodali (2009) utilized a comprehensive list of lean practices and key performance indicators to benchmark two companies against a third, world-class exemplary benchmark (Toyota Motor Corporation). They highlighted the gaps in terms of performance and practices between the two organizations and the benchmark. Other authors have used fuzzy logic approach to determine the gap in leanness performance (Bayou and De Korvin 2008; Vimal and Vinodh 2012; Oleghe and Salonitis 2016a).

- **Impact analysis**

The analysis of the impact of lean is a measure of the effect of lean practices implementation and its improvement on the system. The impact is assessed through the measure of the influence of individual lean practices on themselves and on the leanness level. Anvari et al. (2013) used a Fuzzy Logic approach to emphasize the influence of lean attributes, such as lead time and cost, on leanness, and in the process, identify the critical components to leanness. Oleghe and Salonitis (2016a, b) established the effect of variation in the performance of individual lean practices on the overall leanness trend. Ali and Deif (2014) used a System Dynamics approach to establish the impact of takt time on leanness. Their approach was used to show how adjusting the system's cycle time to align with takt time can be used to improve leanness. Azadeh et al. (2015) used a multi-model approach based on Data Envelopment Analysis, Fuzzy Logic, Decision Making Trial and Evaluation Laboratory and Analytical Hierarchy Process to determine the impact degree of leanness factors on each other as well as on lean strategy. Liang et al. (2015) used the Rooted Arborescence Algorithm model to map the nodal inter-relationships between lean practices in a system. They calculated the degree to which a lean practice supports another lean practice (Liang et al. 2015), and by so doing, were able to prioritize and sequence the lean practices to implement or improve. A number of authors (e.g. Shetty et al. 2010; Stone 2010; Seyedhosseini et al. 2011) have used survey-based and consensus forming methods to identify what practices impact leanness. The results from these methods represent common and empirically established practices and outcomes. Cil and Turkan (2013) used Analytical Network Process modelling approach to measure the relative impact of lean practices on each other and on lean goals. The approach provided information as to which lean practices will achieve which lean goals. Some authors (Ravikumar et al. 2013) have used the impact analysis to establish critical success factors for leanness enhancement.

- **Degree of adoption of lean practices**

The extent to which lean practices are adopted is primarily established using questionnaire-based self-audit tools (Sánchez and Pérez 2001; Doolen and Hacker 2005; Chauhan and Singh 2012; Sezen et al. 2012). Confirming the number of and extent to which lean practices are implemented provides a quick and simple way of checking if a system is lean or not (Goodson 2002). A survey-based analysis has the advantage of being able to establish relationships between the extent to which a company adopts a list of practices and its corporate performance (Sezen et al. 2012). The survey results from the literature is a useful data base that organizations in the early stages of lean transformation of lean can tap into, without having to go through trial and error implementation process. The methodology has been used by Seyedhosseini et al. (2011) to set the criteria for being lean.

- **Benchmarking**

Benchmarking involves contrasting the performances of companies: the information is used to gauge how well a company compares to its peers and those considered as world-class. Benchmarking results enables leanness comparison amongst peers in the same industry (Ray et al. 2006; Bayou and De Korvin 2008), between operations (Ray et al. 2006; Ramachandran and Alagumurthi 2013; Maasouman and Demirli 2015), between manufactured parts (Wan and Chen 2008) and between lean practices (Vinodh and Balaji 2011; Vimal and Vinodh 2013; Anvari et al. 2014; Wagner et al. 2014). It is particularly well suited for indicating ill performing areas (Chhabi et al. 2014). Benchmarking has been obtained using a checklist of lean practices using a world-class manufacturer as the benchmark (Anand and Kodali 2009), by Fuzzy Logic analysis (Bayou and De Korvin 2008; Vinodh and Balaji 2011; Vimal and Vinodh 2013), by Factor Analysis (Ray et al. 2006), by Multi criteria Decision Making methods (Ramachandran and Alagumurthi 2013; Anvari et al. 2014) and by Data Envelopment Analysis (Wan 2006; Wan and Chen 2008).

Selecting an appropriate benchmark for leanness assessment may not be so straightforward. For example, there may be unavailability of benchmarking data to use (Bayou and De Korvin 2008).

- **Scenario analysis**

Scenario analysis is the investigation of the outcomes of various interventions in the system, such as lean improvements. Scenario analysis allows a company estimate the outcome of lean interventions (Ray et al. 2006), optimize lean improvements (Srinivasaraghavan and Allada 2006), verify the impact on leanness from variables that are endogenous and exogenous (Ali and Deif 2016) and perform trade-off analysis amongst competing practices and their improvements (Wan and Chen 2008; Ali and Deif 2014). Some authors have provided approaches that can be used to estimate leanness using 'if-then' fuzzy rules, such that IF a system has 'High Profitability' AND 'High Defect', THEN it will also have 'Medium Availability' (Vinodh et al. 2011). Meanwhile, Ali and Deif (2016) have used their approach to systematically and objectively generate viable scenarios for testing, thereby helping the decision-making process.

Scenario analysis has been predominantly achieved using quantitative tools. Ray et al. (2006) applied Factor Analysis approach, Srinivasaraghavan and Allada (2006) used the Mahalanobis Taguchi Gram Schmidt method, Wan and Chen (2008) used Data Envelopment Analysis, Vinodh et al. (2011) used Fuzzy Logic numbering set, Ali and Deif (2014, 2016) applied System Dynamics methodology.

- **Dynamic analysis**

Dynamic analysis as it relates to the current study is the investigation of the complex interactions that make up a system as well as create changes in the state of the system over time (Sterman 2000). It seeks to investigate the complex interactions between practices (Sterman 2000; Ali and Deif 2014; Oleghe and Salonitis 2016b) and can project leanness into the future (Taleghani et al. 2010; Ali and Deif 2016) thereby supporting strategy formulation. System dynamics simulation modelling has been used by various authors (Taleghani et al. 2010; Ali and Deif 2014, 2016; Oleghe and Salonitis 2016b).

The seven categories identified and shown in Fig. 3, represent the prevalent and common uses of leanness assessment methods in the literature. Table 4 (in Appendix) summarizes the various studies presented in the literature and classifies them as per the seven categories identified.

The way in which past authors have used the leanness assessment methods is indicative of what the methods can be used for. Table 4 has been tallied to provide a more meaningful summary of the trends relating to the uses of leanness assessment methods; Fig. 4 is the chart of these tallies.

Assessing leanness on the basis of quantifying it represents the highest key use of leanness assessment methods (see Fig. 4). About 44% of authors have quantified the level or degree of leanness as a way of assessing lean performance, while 34% have investigated the impact of leanness (Impact Analysis) on the organization (see second highest tally in Fig. 4). Scenario and dynamic Analysis are the least addressed key user benefits. One reason for their low tally is that they have not gained popularity, and it is possible that researchers are yet to appreciate their unique offerings.

A closer inspection of the seven key uses reveals two broad categories: Measurement of Leanness Level and Measurement of Leanness Impact. Measurement of Leanness Level is more concerned with providing an indication of how well the organization is doing in its lean transformation journey. Quantification of Leanness, Gap Analysis, Degree of Adoption and Benchmarking represent similar benefit i.e. measurement of leanness level. These four items constitute roughly 67% of what the methods have been used for. On the other hand, the Measurement of Leanness Impact as the name implies, seeks to provide information about what the likely effect of leanness introduction or its improvement will have

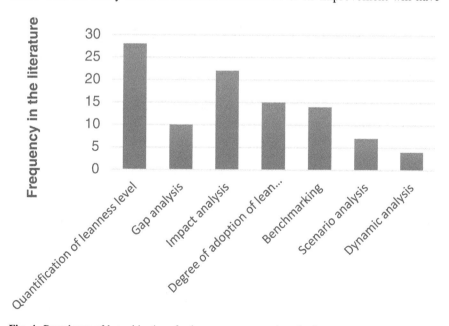

Fig. 4 Prevalence of key objectives for leanness assessment methods

on the system. Impact Analysis, Scenario Analysis and Dynamic Analysis represent 33% of methods used for measurement of leanness impact.

4 Tools and Instruments Used in Leanness Assessment

Various tools and instruments have been used in the leanness assessment methods. Some are generic, others are customized, yet others are modified versions of generic tools. In a number of cases, the tools are used in combination.

Analytical Hierarchy Process (AHP), Analytical Network Process (ANP) and Multi-Criteria Decision-Making Analysis are based on similar concepts and useful where multiple measurement aspects and multiple objectives need to be considered in the assessment process (Nasab et al. 2012; Wong et al. 2014). AHP and ANP help to produce a ranking of elements based on their importance to a common goal such as leanness (Searcy 2009).

Analytical and mathematical equations have been formulated to assess leanness. The formulas are based on a weighted sum of various metrics used for assessing leanness (Toloie-Eshlaghy and Kalantary 2011; Deif et al. 2015).

Artificial Neural Network is often used to find the non-linear relationship in performance data and has been applied to improve the computation of AHP-based techniques in leanness assessment (Nasab et al. 2012) and Fuzzy Logic-based leanness applications (Vimal and Vinodh 2013).

The benefits of benchmarking tool have been provided in the previous section. Balanced Scorecard (BSC) is a performance assessment framework that uses a set of financial and non-financial measures to show an organization's performance trend towards meeting its goals. (Bhasin 2008; Seyedhosseini et al. 2011). Data envelopment analysis (DEA) is a non-parametric mathematical programming technique that supports objective benchmarking analysis (Wan et al. 2007; Wan and Chen 2008).

Decision Making Trial Evaluation Laboratory (DEMATEL) has been used to identify the cause and effect relationships among lean objectives (Seyedhosseini et al. 2011), and to assess the degree of influences that leanness indicators have on each other and on overall leanness Azadeh et al. (2015). Decision Support System is a computer-based Software specifically developed and used for automating the assessment of leanness and enabling quick decision-making process (Wan and Chen 2009; Vinodh and Balaji 2011). Factor analysis is a statistical approach that models the relationships among quantifiable variables, in order to identify variables that cannot be directly measured, such as the leanness of a company (Ray et al. 2006). Fuzzy logic is typically used where impreciseness and vagueness in qualitative data exists, such as data collected using linguistic descriptions of performance (Vinodh and Balaji 2011; Vinodh and Chintha 2011; Vinodh et al. 2011). It has also been used to account for the relative nature of leanness (Bayou and De Korvin 2008) and as a modelling technique to score overall lean performance (Behrouzi and Wong 2011; Pakdil and Leonard 2014). Mahalanobis Taguchi Gram Schmidt

(MTGS) is a statistical based tool that has been used to indicate the degree of abnormality in leanness and identify directions of improvement for a given set of capital constraints (Srinivasaraghavan and Allada 2006).

The functions of questionnaires and self-assessment instruments are straight-forward: to collect and analyse qualitative data in an organized fashion, using yes/no responses (Goodson 2002) and various multi-level grading schemes, (e.g. Hallam 2003a, b; Taj 2005; Vimal and Vinodh 2012). The Rooted Arborescence Algorithm (RAA) maps the nodal inter-relationships between items that make up a system. It has been used to calculate the degree to which a lean practice supports another lean practice (Liang et al. 2015). System dynamics is a simulation mod-elling tool used for investigating the complex inter-relationships that exists within a system, such as the interactions between various lean aspects and the leanness indicators, and has been used by Taleghani et al. (2010), Ali and Deif (2014, 2016), Oleghe and Salonitis (2016b). Value stream mapping, as the name implies, maps the flow of a product, its processes and the associated performance information along the entire value chain, contrasting the current lean state with a future ideal lean state. VSM is a lean improvement tool and it has been used to substantiate the lean state of a system (Wan et al. 2007; Vinodh et al. 2010).

Table 5 (in appendix) has been used to tally the prevalent tools and instruments used in leanness assessment; Fig. 5 is the chart of these tallies. It can be seen from Fig. 5 that questionnaire/self-assessment instrument represents the most prevalent tool used in the literature, constituting the highest frequency with about 28%. This was followed by the use of FL (23%) and AHP/ANP/MCDA (13%). These three sets of tools represent the prevalent tools that have been used in most leanness assessment methods. Meanwhile, majority of the tools (about 63%) were used in only one or two studies, implying their unpopularity in leanness assessment.

While about sixteen different tools and instruments have been used, these tools can be classified under two broad thematic categories: qualitative or quantitative. As observed in the literature, qualitative tools are basically used to collect qualitative types of data such as perceptions of lean performance. Quantitative tools can be described as being analytical in nature: they are used to analyse qualitative or quantitative types of data, and take the form of mathematical expressions. In AHP for example, mathe-matical expressions are used in pairwise comparisons of criteria to establish their individual weights with respect to a common goal (Saaty 2008; Searcy 2009). The self-assessment questionnaire is a qualitative tool, whereas FL or AHP (analytical/quantitative) have been used in quantifying the results from the self-assessment lean audit. Of the sixteen tools identified in the literature, three (benchmarking, BSC and questionnaire/self-assessment) are qualitative tools while others can be classified as quantitative/analytical-type tools. It can be summarized that quantitative/analytical tools are predominantly applied in leanness assessment. This would make sense, since qualitative tools such as the self-assessment questionnaire, on their own, may not be able to reveal the level of leanness, for example.

By combining the information in Tables 4 and 5, the tools/instruments can be ascribed to each generic use of the methods, and this is shown in Table 1. This helps to indicate what tools/instruments can be used for the various applications.

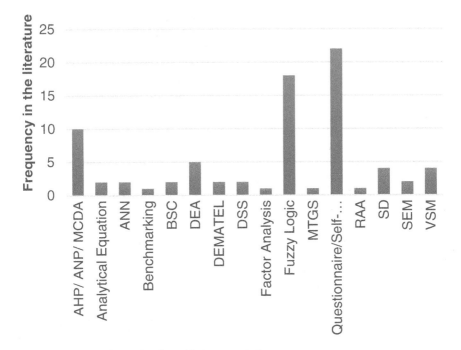

Fig. 5 Prevalence of use of tools and instruments in leanness assessment

Majority of the tools have been used to address at least three key benefits. Only three tools (AHP/ANP/MCDA, Fuzzy Logic and Questionnaires have been used to address five of the key benefits. None of the models has been used to address all seven key benefits. Meanwhile, in Table 2 is the summary of how the tools were applied towards the Measurement of Leanness Level and Measurement of Leanness Impact. From Table 2 it is easy to summarize that most of the tools can be used to measure both the leanness level and leanness impact. However, the tools have not been used by majority of past researchers to achieve both in a single study. In other words, there have been limited use of the tools to simultaneously assess the level of leanness and the impact of leanness. In majority of cases, it has been either one or the other aspect that was addressed.

5 Data Types Used in Leanness Assessment

It would appear that there are varied types of data that have been used in leanness assessment. However, the data types can be classified under two distinct categories: quantitative and qualitative. For the purpose of the review in this chapter and to avoid any ambiguity, reference to quantitative data are those that are directly measurable and are numeric in nature. They are amenable to statistical manipulation

Table 1 Models and instruments of leanness assessment and what they have been used to achieve

Instruments	Uses						
	Quantification of leanness level	Gap analysis	Impact analysis	Degree of adoption of lean practices	Benchmarking	Scenario analysis	Dynamic analysis
AHP/ANP/MCDM	X		X	X	X	X	
Analytical equation	X						
ANN	X				X		
Benchmarking	X	X		X	X		
BSC			X				
DEA	X	X	X		X		
DEMATEL			X				
DSS	X						
Factor analysis	X				X	X	
Fuzzy Logic	X	X	X		X	X	
MTGS	X	X			X	X	
Questionnaire/Self-assessment	X	X	X	X	X		
RAA					X		
SD	X		X			X	X
SEM			X	X			
VSM	X		X				

Table 2 Leanness assessment instruments and how they have been used to address the measurement and impact of leanness

Instruments	Uses	
	Measurement of leanness level	Measurement of leanness impact
AHP/ANP/MCDM	X	X
Analytical equation	X	
ANN	X	
Benchmarking	X	
BSC	X	
DEA	X	X
DEMATEL		X
DSS	X	
Factor analysis	X	X
Fuzzy Logic	X	X
MTGS	X	X
Questionnaire/ Self-assessment	X	X
RAA	X	
SD	X	X
SEM	X	X
VSM	X	X

in their raw and original form. Qualitative data are those types of data that are not directly measurable or quantifiable.

There are many generic quantitative data types associated with leanness assessment such as lead time, defect rate, setup time, changeover time, work-in-process quantity, mean time to machine breakdown, mean time to machine repair, overall equipment effectiveness, first time through capability and dock-to-dock time. Other types of quantitative data have been used in leanness assessment. Behrouzi and Wong (2011) used a number of orders delivered late and lead time to represent JIT lean aspect. Srinivasaraghavan and Allada (2006) used a number of kaizen events in their leanness assessment method to represent Continuous Improvement programmes. Bayou and De Korvin (2008) used relative change in inventories over time to represent Kaizen, while Pakdil and Leonard (2014) used space productivity (production throughput per square foot of space) to represent process related lean dimension.

Qualitative data types in leanness assessment can also be referred to as soft metrics, as they capture information on such items like lean culture, leadership commitment to continuous improvement, extent of the use of lean tools and adherence level to lean. They are not directly measurable, rather they are measured using descriptive words such as 'Worst', 'Very poor', 'Poor', 'Fair', 'Good', 'Very good' and 'Excellent' (see for example Zanjirchi et al. 2010; Vimal and Vinodh 2012; Chhabi et al. 2014) or assigned points on a rated scale such as a Likert scale

(Hallam 2003a, b; Seyedhosseini et al. 2011; Wong et al. 2014). A 100 points scale has also been used to collect qualitative data (Taj 2005; Singh et al. 2010; Anvari et al. 2013). Information about lean performance for these types of data are typically questionnaire generated.

Qualitative data types have been used to assess the degree or extent of lean adoption. The LESAT was used by Hallam (2003a, b) to assess the degree of lean adoption in the aerospace industry. Taj (2005) used the Strategos LAT to assess the level of lean adoption in Chinese Hi-Tech industries using a 100-point scale. Anand and Kodali (2009) used a checklist of lean practices adopted in Toyota Motor Corporation to benchmark the number of adopted lean practices in two organizations. Qualitative leanness indicators have also been used to assess the criteria and enablers for being lean (Chauhan and Singh 2012; Vinodh and Joy 2012; Anvari et al. 2013; Cil and Turkan 2013; Gupta et al. 2013; Chhabi et al. 2014; Wong et al. 2014). Table 6 has been used to indicate the prevalence of each data type in the literature.

From Table 6 it can be seen that qualitative types of data have been predominantly used in the literature when compared to the quantitative types of data. This trend may not be unconnected with the difficulty in obtaining quantitative types of data. In other instances, the data may not be accurate or readily available (Ray et al. 2006), while in others, the data may be outdated to provide information about current leanness (Bayou and De Korvin 2008; Searcy 2009). Qualitative types of data can be easily captured using a self-assessment questionnaire tool. For organizations, qualitative types of data are less sensitive and less proprietary than quantitative data, hence their popularity in majority of the methods that have been advanced in the literature.

While the use of qualitative types of data have been predominant in leanness assessment methods, the use of mixed-data types (i.e. quantitative and qualitative) in a unified leanness assessment has been rarely applied. Of the 64 studies that were set apart for the survey, only two (Pakdil and Leonard 2014; Oleghe and Salonitis 2016a, b) have used mixed-data type in their leanness assessment. From the findings, it can be taken that the data used in majority of the methods are either distinctively quantitative or qualitative.

Table 3 has been used to sort out which data types have been used to achieve the different leanness assessment objectives. It can be seen from Table 3 that both data types have been used in achieving the objectives of the leanness assessment. It was found that only qualitative data types are used when the degree of adoption of lean practices is the objective of the assessment.

6 Conceptual Framework for Leanness Assessment and Discussion

Conceptual maps are needed in the design of conceptual frameworks, which can be used to develop an appropriate leanness assessment method for specific situations. While Fig. 1 (in the Introduction section) describes the basic format for leanness

Table 3 Uses of leanness assessment and the data types

Key objectives	Data type	
	Qualitative data	Quantitative data
Quantification of leanness level	X	X
Gap analysis	X	X
Impact analysis	X	X
Degree of adoption of lean practices	X	
Benchmarking	X	X
Scenario analysis	X	X
Dynamic analysis	X	X

assessment methods, the trends and patterns that have been revealed in the current study can be used to provide better clarity and understanding in the design and use of the methods. By revisiting Fig. 1 and interjecting the revealed patterns in the literature, a conceptual map for leanness assessment method design (Fig. 6) can be advanced. In this conceptual map, an additional level in the hierarchy has been added at the beginning (top) of the framework to specify the objectives of the leanness assessment as compared to the one depicted in Fig. 1.

The conceptual model shown in Fig. 6 has been presented as part of the DMAIC (Define, Measure, Analysis, Improve and Control) circle for improvement. Obviously, the leanness assessment can be mapped only to the Define, Measure and Analysis phases. Analysis stops with the analysis of the leanness levels per LM aspect investigated. This needs to be complemented by setting up plans for improvement. Figure 6 presents high-level information with regards the Improve and Control phases.

Advancing the conceptual frameworks that are possible from the conceptual map is beyond the aim of the current study. This should be an area where other researchers can advance the work. An example of a conceptual framework could be one in which the objective of the leanness assessment is for benchmarking purposes, representing the first level in the hierarchy. The aspects of LM to be considered for the assessment (representing the second level in the hierarchy) could be based either on the characteristics of the organization considering the assessment or on the objectives of the assessment, or both, depending on choice. Ideally, a comprehensive set of practices relevant to the organization type should be adopted for the assessment. On the third level of the hierarchy is the data type. The authors of the current chapter advocate for a mixed-data type leanness assessment. If a qualitative data type is to be used, the self-assessment questionnaire should be the instrument of choice for initial data collection. The DEA/AHP (Anvari et al. 2014; Azadeh et al. 2015) can be used in the final analysis of the data obtained using the questionnaire. In other words, the fourth level will indicate multiple tools- questionnaire, DEA and AHP. If a quantitative data type is considered, the DEA (Wan 2006; Wan and Chen 2008) the FA (Ray et al. 2006) or FL (Bayou and De Korvin 2008) tools can be applied. The fifth level is simply the outcome of the leanness

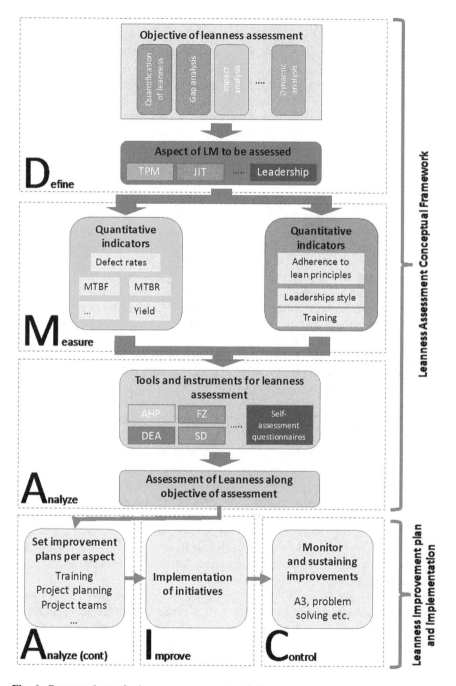

Fig. 6 Conceptual map for leanness assessment methods

assessment which should correspond with the initial objectives of the assessment. Other conceptual frameworks can be gleaned from the current study key findings.

The conceptual framework should integrate the key features of leanness assessment, defining the tools and data that are useful for specific leanness assessment objectives. The framework should guide adopters towards a structured approach to follow and prevent the random selection of tools/instruments as well as data types.

Others (Wahab et al. 2013) have developed a conceptual model of leanness measurement for the manufacturing industry, however, their conceptual model focused only on two levels namely dimensions (aspects of LM) and factors (indicators of leanness). The conceptual map described above contains multiple levels that cover all parts of the assessment. Conceptual maps have also been generated for LM practices selection (Shah and Ward 2003; Pavnaskar et al. 2003; Mirdad and Eseonu 2015); there is yet a conceptual map for leanness assessment method selection.

The quantification of leanness level has occupied majority of past research efforts such that the leanness assessment methods have been focused in this area. The current state of saturation suggests that it is a mature area and there is no need for further methods focusing on this aspect. It also suggests that advances have been made to the point where the current methods are sufficient enough for most situations. So rather than add to the body of knowledge, researchers can focus their efforts on testing the reliability and robustness of the information, the tools and the data requirements for each method.

The measurement of leanness appears to be well investigated by past researchers, suggesting a saturated field, where additions would be of little value. In short, the advanced methods have fully covered this aspect. Whereas, the investigation of the impact of leanness is less researched, implying that more methods are required to advance the knowledge in this area. Of the methods that have been advanced in investigating the impact of leanness, majority are subjective-based tools that rely on human judgement and perceptions about the relationship between leanness and organizational performance. There is a paucity of methods to provide an objective assessment of how lean impacts organizational performance. In fact, Hopp et al. (2007) and Marvel and Standridge (2009) express that achieving the benefits of lean has often been approached using an intuitive method of improvement. While majority of the methods can be used to measure leanness and also indicate the likely impact of leanness, there have been limited methods advanced to accomplish both aspects in a single assessment. This would certainly be of greater value than addressing either one separately.

In terms of types of data used in the methods, it can be seen that there are more methods that use qualitative types of data. This may due to the difficulty in collecting accurate and reliable quantitative data. It could also be due to organizations being protective of their figures. In other instances, the quantitative data simply does not exist (Ray et al. 2006). Both types of data have their strengths and weaknesses, so it means that the leanness assessment methods would also exhibit the strengths and weaknesses of each data type. However, there is an extreme paucity of methods that use mixed data i.e. both qualitative and quantitative in the same assessment. This indicates a shortfall in methods that use mixed-data types to combine the strengths of each data type and overcome their respective weaknesses. In addition, Pakdil and Leonard (2014) suggest that it is better to use both types as they may individually indicate different directions of performance, thereby providing misleading information. If used in a combined manner, they individually act as a check for the other, thereby improving the confidence in the assessment result. Can a single leanness assessment method achieve this? The answer is yes, as Pakdil and Leonard (2014) have used FL, even though they advanced two separate LATs in their advanced method. In their approach, they used 'hard' quantitative measures to develop one LAT and 'soft' qualitative measures for the other. Can a single tool be used as an all-inclusive method, where both qualitative and quantitative data are homogenized in a holistic method? The answer to this question is also yes. For example, Zahn et al. (1998) have used an SD approach to justify the investments in a manufacturing plant, on the basis of hard and soft decision criteria. Taleghani et al. (2010), Ali and Deif (2014, 2016), Oleghe and Salonitis (2016a, b) have all used SD tool in their leanness assessment approach. There have been very few applications of SD in the literature, and this suggests an area where more research efforts may be directed to fully explore its potentials.

Seven broad areas were identified, which represent the generic uses of the methods. It was revealed that majority of the tools used in the methods have been used to address most areas. However, there have been limited research efforts undertaken whereby the advance methods address most if not all of the areas. What this means is that while there are several uses and applications, none of the advanced methods has been used to address the multiple areas, simultaneously. Leanness assessment methods would be of greater benefit if they address many issues rather than a few set of issues.

Trends in past data can be used to guide future behaviours. The prevalent trends revealed in the current study are useful for future researchers. For example, a comprehensive and exhaustive set of tools and instruments have been revealed, from which future researchers can draw from. In addition, there are commonalities that can be gleaned from the current study. These commonalities are useful in creating paths that link data types with tools/instruments and with uses of the methods. The trend analysis revealed from the literature survey provides guidance as to which model can be used for a specific purpose. For example, scenario and dynamic analyses have been approached using a system dynamics tool, while quantification of leanness level has been approached using virtually all the prevalent tools. Assessing the Degree of Adoption of lean practices can, for example, be undertaken using questionnaire/self-assessment instruments. These and many more linkages are derivable from the current document.

Industrial implication is a consequence of academic implications. Wherever there is an abundance of methods or applications, there is the likelihood of advances in knowledge. Where there are advances in knowledge, industry benefits as the eventual users of the methods. Areas where there has been an overemphasis suggest areas where practitioners can be confident of the reliability of the method, as being tried and tested. Vice versa is the case where there is a dearth of methods or knowledge. In other words, methods that incorporate self-assessment questionnaire tools and fuzzy logic, methods that are based on qualitative data types and methods that quantify or measure leanness level are those in which industry will have confidence using. Meanwhile, methods that utilize less prevalent tools such as the Rooted Arborescence Algorithm (RAA) and the Mahalanobis Taguchi Gram Schmidt (MTGS) will not likely be tried out. So while methods based on these tools may be advantageous, industry may probably not benefit from its use since a practitioner will most likely choose a method that has been tried and tested.

One aspect that calls for concern is the paucity of methods that measure or investigate the impact of leanness; methods that are not based on subjective human perception or judgment. With the current trend, it means that industry practitioners will approach leanness improvement on the bases of gut feeling and intuition. In other cases, they will use information from documented empirical studies, and assume that the findings in those empirical studies will be applicable or replicable in their own case. Meanwhile, those involved with lean introduction or its improvement should be able to show their clients or organizations that lean will actually improve company performance, and be specific on the magnitude of change. In the absence of this, the organization is less likely to adopt or be committed to improving lean.

7 Conclusions

This study set out to survey the literature for leanness assessment methods. The intention was to reveal common themes and trends as well as gaps, to provide directions where future advancements can be made in the methods. The survey results generated various trends through which conceptual frameworks can be developed. The gaps indicate paucity in methods using mixed (quantitative and qualitative) data type. The gaps also reveal the paucity of methods that provide an unambiguous evidence of the impact of leanness. The scope of work did not include an empirical survey: something that will establish what is really obtainable in practice. For example, the Value Stream Mapping tool is well established in lean organizations, whereas it is rarely used in the literature for leanness assessment. On the other hand, the self-assessment instruments are prevalent in the literature, but there are no empirical studies to indicate their prevalence in practice. It is also possible that many organizations lack the expertise and the commitment to generate a leanness assessment methodology on the one hand and to apply it on the other. The empirical knowledge is limited. A research effort may be one that seeks to carry out an empirical survey to establish the prevalence of each instrument as well as the typical uses.

Appendix

See Tables 4, 5 and 6.

Table 4 The prevalence of the generic uses of leanness assessment methods

S/No.	Author(s)	Year	Quantification of leanness level	Gap analysis	Impact analysis	Degree of adoption of lean practices	Benchmarking	Scenario analysis	Dynamic analysis
1	Sánchez and Pérez	2001			X	X			
2	Goodson	2002				X			
3	Soriano-Meier and Forrester	2002				X			
4	Nightingale and Mize	2002		X					
5	Kumar and Thomas	2002				X			
6	Hallam	2003a, b				X			
7	Doolen and Hacker	2005				X			
8	Taj	2005	X	X			X		
9	Srinivasaraghavan and Allada	2006	X	X			X	X	
10	Wan	2006	X		X		X		
11	Ray et al.	2006	X				X		
12	Wan et al.	2007	X	X					
13	Bhasin	2008			X				
14	Wan and Chen	2008	X		X		X		
15	Bayou and De Korvin	2008	X				X		
16	Ihezie and Hargrove	2009	X	X			X		
17	Anand and Kodali	2009		X		X	X		
18	Wan and Chen	2009		X					
19	Shetty et al.	2010	X			X			

(continued)

Table 4 (continued)

S/No.	Author(s)	Year	Quantification of leanness level	Gap analysis	Impact analysis	Degree of adoption of lean practices	Benchmarking	Scenario analysis	Dynamic analysis
20	Singh et al.	2010	X						
21	Taleghani et al.	2010	X	X	X			X	X
22	Vinodh et al.	2010			X				
23	Zanjirchi et al.	2010	X						
24	Anvari et al.	2011			X				
25	Behrouzi and Wong	2011	X						
26	Seyedhosseini et al.	2011			X				
27	Toloie-Eshlaghy and Kalantary	2011	X				X		
28	Vinodh and Balaji	2011	X						
29	Vinodh and Chintha	2011	X						
30	Vinodh et al.	2011			X			X	
31	Chauhan and Singh	2012				X			
32	Deif	2012			X				
33	Nasab et al.	2012						X	
34	Sezen et al.	2012				X			
35	Stone	2012			X				
36	Vinodh and Joy	2012				X			
37	Vimal and Vinodh	2012		X				X	
38	Vinodh and Vimal	2012	X						
39	Alemi and Akram	2013	X						
40	Anvari et al.	2013			X				

(continued)

Table 4 (continued)

S/No.	Author(s)	Year	Quantification of leanness level	Gap analysis	Impact analysis	Degree of adoption of lean practices	Benchmarking	Scenario analysis	Dynamic analysis
41	Cil and Turkan	2013			X				
42	Gupta et al.	2013			X				
43	Ravikumar et al.	2013			X				
44	Ramachandran and Alagumurthi	2013	X						
45	Vimal and Vinodh	2013	X				X		
46	Ali and Deif	2014	X		X			X	X
47	Berlec et al.	2014	X						
48	Anvari et al.	2014			X		X		
49	Chhabi et al.	2014			X		X		
50	Giri et al.	2014			X				
51	Hallam and Keating	2014				X			
52	Pruthvi and Sreenivasa	2014		X					
53	Wong et al.	2014	X						
54	Wagner et al.	2014				X			
55	Pakdil and Leonard	2014	X						
56	Azadeh et al.	2015			X				
57	Deif et al.	2015	X						
58	Maasouman and Demirli	2015	X				X		
59	Liang et al.	2015					X		

(continued)

Table 4 (continued)

S/No.	Author(s)	Year	Quantification of leanness level	Gap analysis	Impact analysis	Degree of adoption of lean practices	Benchmarking	Scenario analysis	Dynamic analysis
60	Schröders and Cruz-Machado	2015				X			
61	Susilawati et al.	2015				X			
62	Ali and Deif	2016	X		X			X	X
63	Oleghe and Salonitis	2016a	X		X				
64	Oleghe and Salonitis	2016b							X
Tally			**28**	**10**	**22**	**15**	**14**	**7**	**4**

Table 5 Tools and instruments that have used in leanness assessment

S/No.	Authors	Year	AHP/ANP/MCDA	Analytical equation	ANN	Benchmarking	BSC	DEA	DEMATEL	DSS
1	Sánchez and Pérez	2001								
2	Goodson	2002								
3	Soriano-Meier and Forrester	2002								
4	Nightingale and Mize	2002								
5	Kumar and Thomas	2002								
6	Hallam	2003a, b								
7	Doolen and Hacker	2005								
8	Taj	2005								
9	Srinivasaraghavan and Allada	2006								
10	Wan	2006						X		
11	Ray et al.	2006								
12	Wan et al.	2007						X		
13	Bhasin	2008					X			
14	Wan and Chen	2008						X		
15	Bayou and De Korvin	2008								
16	Ihezie and Hargrove	2009								
17	Anand and Kodali	2009				X				
18	Wan and Chen	2009								X
19	Shetty et al.	2010								
20	Singh et al.	2010								

(continued)

Table 5 (continued)

S/No.	Authors	Year	AHP/ANP/MCDA	Analytical equation	ANN	Benchmarking	BSC	DEA	DEMATEL	DSS
21	Taleghani et al.	2010								
22	Vinodh et al.	2010								
23	Zanjirchi et al.	2010								
24	Anvari et al.	2011	X							
25	Behrouzi and Wong	2011								
26	Seyedhosseini et al.	2011					X		X	
27	Toloie-Eshlaghy and Kalantary	2011	X	X						
28	Vinodh and Balaji	2011								X
29	Vinodh and Chintha	2011								
30	Vinodh et al.	2011								
31	Chauhan and Singh	2012	X							
32	Deif	2012								
33	Nasab et al.	2012	X		X					
34	Sezen et al.	2012								
35	Stone	2012								
36	Vinodh and Joy	2012								
37	Vimal and Vinodh	2012								
38	Vinodh and Vimal	2012								
39	Alemi and Akram	2013								
40	Anvari et al.	2013								
41	Cil and Turkan	2013	X							

(continued)

Table 5 (continued)

S/No.	Authors	Year	AHP/ANP/MCDA	Analytical equation	ANN	Benchmarking	BSC	DEA	DEMATEL	DSS
42	Gupta et al.	2013								
43	Ravikumar et al.	2013	X							
44	Ramachandran and Alagumurthi	2013	X							
45	Vimal and Vinodh	2013			X					
46	Ali and Deif	2014								
47	Berlec et al.	2014								
48	Anvari et al.	2014	X					X		
49	Chhabi et al.	2014								
50	Giri et al.	2014								
51	Hallam and Keating	2014								
52	Pruthvi and Sreenivasa	2014								
53	Wong et al.	2014	X							
54	Wagner et al.	2014								
55	Pakdil and Leonard	2014								
56	Azadeh et al.	2015	X					X	X	
57	Deif et al.	2015		X						
58	Maasouman and Demirli	2015								
59	Liang et al.	2015								
60	Schröders and Cruz-Machado	2015								

(continued)

Table 5 (continued)

S/No.	Authors	Year	AHP/ANP/MCDA	Analytical equation	ANN	Benchmarking	BSC	DEA	DEMATEL	DSS
61	Susilawati et al.	2015								
62	Ali and Deif	2016								
63	Oleghe and Salonitis	2016a								
64	Oleghe and Salonitis	2016b								
Tally			10	2	2	1	2	5	2	2

S/No.	Authors	Year	Factor analysis	Fuzzy logic	MTGS	Questionnaire/Self-assessment	RAA	SD	SEM	VSM
1	Sánchez and Pérez	2001				X				
2	Goodson	2002				X				
3	Soriano-Meier and Forrester	2002				X				
4	Nightingale and Mize	2002				X				
5	Kumar and Thomas	2002				X				
6	Hallam	2003a, b				X				
7	Doolen and Hacker	2005				X				
8	Taj	2005				X				
9	Srinivasaraghavan and Allada	2006			X					
10	Wan	2006								
11	Ray et al.	2006	X							X
12	Wan et al.	2007								
13	Bhasin	2008								

(continued)

Table 5 (continued)

S/No.	Authors	Year	Factor analysis	Fuzzy logic	MTGS	Questionnaire/Self-assessment	RAA	SD	SEM	VSM
14	Wan and Chen	2008								
15	Bayou and De Korvin	2008		X						
16	Ihezie and Hargrove	2009				X				
17	Anand and Kodali	2009								
18	Wan and Chen	2009								
19	Shetty et al.	2010				X				
20	Singh et al.	2010		X						
21	Taleghani et al.	2010						X		
22	Vinodh et al.	2010								X
23	Zanjirchi et al.	2010		X						
24	Anvari et al.	2011								
25	Behrouzi and Wong	2011		X						
26	Seyedhosseini et al.	2011								
27	Toloie-Eshlaghy and Kalantary	2011								
28	Vinodh and Balaji	2011		X						
29	Vinodh and Chintha	2011		X						
30	Vinodh et al.	2011		X						
31	Chauhan and Singh	2012				X				
32	Deif	2012								X
33	Nasab et al.	2012								
34	Sezen et al.	2012				X				

(continued)

Table 5 (continued)

S/No.	Authors	Year	Factor analysis	Fuzzy logic	MTGS	Questionnaire/Self-assessment	RAA	SD	SEM	VSM
35	Stone	2012				X				
36	Vinodh and Joy	2012				X			X	
37	Vimal and Vinodh	2012		X						
38	Vinodh and Vimal	2012		X						
39	Alemi and Akram	2013		X						
40	Anvari et al.	2013		X						
41	Cil and Turkan	2013								
42	Gupta et al.	2013				X				
43	Ravikumar et al.	2013								
44	Ramachandran and Alagumurthi									
45	Vimal and Vinodh	2013		X						
46	Ali and Deif	2014						X		
47	Berlec et al.	2014								X
48	Anvari et al.	2014				X				
49	Chhabi et al.	2014		X						
50	Giri et al.	2014		X						
51	Hallam and Keating	2014				X				
52	Pruthvi and Sreenivasa	2014				X				
53	Wong et al.	2014								
54	Wagner et al.	2014				X				
55	Pakdil and Leonard	2014		X						
56	Azadeh et al.	2015		X						

(continued)

Table 5 (continued)

S/No.	Authors	Year	Factor analysis	Fuzzy logic	MTGS	Questionnaire/Self-assessment	RAA	SD	SEM	VSM
57	Deif et al.	2015								
58	Maasouman and Demirli	2015				X				
59	Liang et al.	2015					X			
60	Schröders and Cruz-Machado	2015				X				
61	Susilawati et al.	2015		X		X				
62	Ali and Deif	2016						X		
63	Oleghe and Salonitis	2016a		X						
64	Oleghe and Salonitis	2016b						X		
Tally			1	18	1	22	1	4	2	4

Table 6 Prevalence of data types used in leanness assessment

S/No.	Authors	Year	Quantitative data type	Qualitative data type	Mixed-data
1	Sánchez and Pérez	2001		X	
2	Goodson	2002		X	
3	Soriano-Meier and Forrester	2002		X	
4	Nightingale and Mize	2002		X	
5	Kumar and Thomas	2002		X	
6	Hallam	2003a, b		X	
7	Doolen and Hacker	2005		X	
8	Taj	2005		X	
9	Srinivasaraghavan and Allada	2006	X		
10	Wan	2006	X		
11	Ray et al.	2006	X		
12	Wan et al.	2007	X		
13	Bhasin	2008		X	
14	Wan and Chen	2008	X		
15	Bayou and De Korvin	2008	X		
16	Ihezie and Hargrove	2009		X	
17	Anand and Kodali	2009		X	
18	Wan and Chen	2009	X		
19	Shetty et al.	2010		X	
20	Singh et al.	2010		X	
21	Taleghani et al.	2010	X		
22	Vinodh et al.	2010		X	
23	Zanjirchi et al.	2010		X	
24	Anvari et al.	2011		X	
25	Behrouzi and Wong	2011	X		
26	Seyedhosseini et al.	2011		X	
27	Toloie-Eshlaghy and Kalantary	2011		X	
28	Vinodh and Balaji	2011		X	
29	Vinodh and Chintha	2011		X	
30	Vinodh et al.	2011		X	
31	Chauhan and Singh	2012		X	
32	Deif	2012	X		
33	Nasab et al.	2012		X	
34	Sezen et al.	2012		X	
35	Stone	2012		X	
36	Vinodh and Joy	2012		X	

(continued)

Table 6 (continued)

S/No.	Authors	Year	Quantitative data type	Qualitative data type	Mixed-data
37	Vimal and Vinodh	2012		X	
38	Vinodh and Vimal	2012		X	
39	Alemi and Akram	2013		X	
40	Anvari et al.	2013		X	
41	Cil and Turkan	2013		X	
42	Gupta et al.	2013		X	
43	Ravikumar et al.	2013		X	
44	Ramachandran and Alagumurthi	2013		X	
45	Vimal and Vinodh	2013		X	
46	Ali and Deif	2014		X	
47	Berlec et al.	2014	X		
48	Anvari et al.	2014		X	
49	Chhabi et al.	2014		X	
50	Giri et al.	2014		X	
51	Hallam and Keating	2014		X	
52	Pruthvi and Sreenivasa	2014		X	
53	Wong et al.	2014		X	
54	Wagner et al.	2014		X	
55	Pakdil and Leonard	2014	X	X	X
56	Azadeh et al.	2015		X	
57	Deif et al.	2015	X		
58	Maasouman and Demirli	2015		X	
59	Liang et al.	2015		X	
60	Schröders and Cruz-Machado	2015		X	
61	Susilawati et al.	2015		X	
62	Ali and Deif	2016	X		
63	Oleghe and Salonitis	2016a	X		
64	Oleghe and Salonitis	2016b	X	X	X
Tally			**16**	**50**	**2**

References

Alemi, M. A., & Akram, R. (2013). Measuring the leanness of manufacturing systems by using fuzzy TOPSIS: A case study of the Parizan Sanat company. *South African Journal of Industrial Engineering, 24*(3), 166–174.

Ali, R., & Deif, A. (2014). Dynamic lean assessment for takt time implementation. *Procedia CIRP, 17*, 577–581.

Ali, R., & Deif, A. (2016). Assessing leanness level with demand dynamics in a multi-stage production system. *Journal of Manufacturing Technology Management, 27*(5), 614–639.

Anand, G., & Kodali, R. (2009). Application of benchmarking for assessing the lean manufacturing implementation. *Benchmarking: An International Journal, 16*(2), 274–308.

Anvari, A., et al. (2011). A group AHP-based tool to evaluate effective factors toward leanness in automotive industries. *Journal of Applied Sciences, 11*(17), 3142–3151.

Anvari, A., et al. (2014). An integrated design methodology based on the use of group AHP-DEA approach for measuring lean tools efficiency with undesirable output. *International Journal of Advanced Manufacturing Technology, 70*, 2169–2186.

Anvari, A., Zulkifli, N., & Yussuf, R. (2013). A dynamic modeling to measure lean performance within lean attributes. *International Journal of Advanced Manufacturing Technology, 66*, 663–677.

Association for Manufacturing Excellence. (2016). AME Lean Assessment Tool. http://www.ame.org/ame-lean-assessment-tool. Assessed April 2016.

Azadeh, A., Zarrin, M., Abdollahi, M., Noury, S., & Farahmand, S. (2015). Leanness assessment and optimization by fuzzy cognitive map and multivariate analysis. *Expert Systems with Applications, 42*, 6050–6064.

Bayou, M. E., & De Korvin, A. (2008). Measuring the leanness of manufacturing systems—A case study of Ford Motor Company and General Motors. *Journal of Engineering and Technology Management, 25*(4), 287–304.

Behrouzi, F., & Wong, K. Y. (2011). Lean performance evaluation of manufacturing systems: A dynamic and innovative approach. *Procedia Computer Science, 3*, 388–395.

Berlec, T., Jus, G., Starbek, M., & Kusar, J. (2014). Leanness index of a process chain. *Technics Technologies Education Management, 9*(3), 552–563.

Bhasin, S. (2008). Lean and performance measurement. *Journal of Manufacturing Technology Management, 19*(5), 670–684.

Bonavia, T., & Marin, J. A. (2006). An empirical study of lean production in the ceramic tile industry in Spain. *International Journal of Operations & Production Management, 26*(5), 505–531.

Chauhan, G., & Singh, T. P. (2012). Measuring parameters of lean manufacturing realization. *Measuring Business Excellence, 16*(3), 57–71.

Chhabi, R. M., Datta, S., & Mahapatra, S. S. (2014). Leanness estimation procedural hierarchy using interval-valued fuzzy sets (IVFS). *Benchmarking: An International Journal, 21*(2), 150–183.

Cil, I., & Turkan, Y. S. (2013). An AHP-based assessment model for lean enterprise transformation. *International Journal of Advanced Manufacturing Technology, 64*, 1113–1130.

Deif, A. (2012). Assessing lean systems using variability mapping. *Procedia CIRP, 3*, 2–7.

Deif, A., Janfawi, A. W., & Ali, R. (2015). An integrated metric to assess leanness level based on efficiency, flow and variation". *Journal of Supply Chain and Operations Management, 13*(1), 44–57.

Doolen, T. L., & Hacker, M. E. (2005). A review of lean assessment in organizations: An exploratory study of lean practices by electronics manufacturers. *Journal of Manufacturing Systems, 24*(1), 55–67.

Gemba Academy LLC. (2010). Lean Enterprise Assessment. Available online www.GembaAcademy.com. Accessed June 2015.

Giri, D., Bangar, A., Dubey, V. K., & Giri, D. (2014). Enhancing leanness in manufacturing process of small scale industry using Fuzzy QFD approach. *The International Journal of Science and Technoledge, 2*(7), 102–108.

Goodson, R. E. (2002). Read a plant-fast. *Harvard Business Review, 80*(5), 105–113.

Gupta, V., Acharya, P., & Patwardhan, M. (2013). A strategic and operational approach to assess the lean performance in radial tyre manufacturing in India. *International Journal of Productivity and Performance Management, 62*(6), 634–651.

Hallam, C. R. (2003a). *Lean enterprise self-assessment as a leading indicator of accelerating transformation in the aerospace industry* (Ph.D. Thesis). Massachusetts Institute of Technology.

Hallam, C. R. (2003b). LESAT facilitation: The steps to organizing and running a lean enterprise self-assessment. *LESAT Facilitator Workshop, 27 March.* https://dspace.mit.edu.

Hallam, C., & Keating, J. (2014). Company self-assessment of lean enterprise maturity in the aerospace industry. *Journal of Enterprise Transformation, 4*(1), 51–71.

Hopp, W. J., Iravani, S. M., & Shou, B. (2007). A diagnostic tree for improving production line performance. *Production and Operations Management, 16*(1), 77–92.

Ihezie, D., & Hargrove, S. (2009). Applying lean assessment tools at a Maryland manufacturing company. American Society for Engineering Education (ASEE). https://www.asee.org/documents/sections/middleatlantic/spring-2009/Applying-Lean-Assessment-Tools-at-a-Maryland-Manufacturing-Company.pdf. Accessed 22 Oct 2014.

Kumar, A., & Thomas, S. (2002). A software tool for screening analysis of lean practices. *Environmental Progress, 21*(3).

Lean Advancement Initiative. (2012). LAI Enterprise Self-Assessment Tool (LESAT) Version. 2.0.

Liang, Y., Shan, S., Qiao, L., & Lei, Y. (2015). Criteria for lean practice selection: development an assessment tool using the rooted arborescence. In *IEEE International Conference on Automation Science and Engineering (CASE) in Gothenburg, Sweden, 24–28 August 2015.*

Maasouman, M., & Demirli, K. (2015). Assessment of lean maturity level in manufacturing cells. *IFAC-PapersOnLine, 48*(3), 1876–1881.

Mahfouz, A. (2011). *An integrated framework to assess leanness performance in distribution centres* (Ph.D. Thesis). School of Management, Dublin Institute of Technology.

Marvel, J. H., & Standridge, C. R. (2009). Simulation-enhanced lean design process. *Journal of Industrial Engineering and Management, 2*(1), 90–113.

Mirdad, W. K., & Eseonu, C. I. (2015). A conceptual map of the lean nomenclature/comparing expert classification to the lean literature. *Engineering Management Journal, 27*(4), 188–202.

Nasab, H. H., Aliheidari bioki, T., & Zare, H. K. (2012). Finding a probabilistic approach to analyze lean manufacturing. *Journal of Cleaner Production, 29–30,* 73–81.

Nightingale, D. J., & Mize, J. H. (2002). Development of a lean enterprise transformation maturity model. *Information Knowledge Systems Management, 3,* pp. 15–30.

Oleghe, O., & Salonitis, K. (2015). Improving the efficacy of the lean index through the quantification of qualitative lean metrics. *Procedia CIRP, 37,* 42–47.

Oleghe, O., & Salonitis, K. (2016a). Variation modelling of lean manufacturing performance using fuzzy logic based quantitative lean index. *Procedia CIRP, 41,* 608–613.

Oleghe, O., & Salonitis, K. (2016b). A lean assessment tool based on systems dynamics. *Procedia CIRP, 50,* 106–111.

Pakdil, F., & Leonard, K. M. (2014). Criteria for a lean organization: Development of a lean assessment tool. *International Journal of Production Research, 52*(15), 4589–4607.

Pavnaskar, S. J., Gershenson, J. K., & Jambekar, A. B. (2003). Classification scheme for lean manufacturing tools. *International Journal of Production Research, 41*(13), 3075–3090.

Perkins, L., Abdimomunova, L., Valerdi, R., Shields, T., & Nightingale, D. (2010). Insights from enterprise assessment/How to analyze LESAT results for enterprise transformation. *Information Knowledge Systems Management, 9,* 153–174.

Pruthvi, H. M., & Sreenivasa, C. G. (2014). Quantification of leanness in a textile industry. *International Journal of Computational Engineering Research, 4*(3), 58–62.

Ramachandran, L., & Alagumurthi, N. (2013). Application of key performance indicators in a leather and shoe industry for leanness analysis using multicriteria approach: A pre implementation study. *International Journal of Advance Industrial Engineering, 1*(2), 43–47.

Ravikumar, M., Marimuthu, K., Parthiban, P., & Abdul, Zubar H. (2013). Leanness evaluation in 6 manufacturing MSMEs using AHP and SEM techniques. *International Journal of Mechanical and Mechtronics Engineering, 13*(6), 29–36.

Ray, C. D., Zuo, X., Micheal, J. H., & Wiedenbeck, J. K. (2006). The lean index: operational lean metrics for the wood products industry. *Wood and Fiber Science, 38*(2), 238–255.

Rymaszewska, A. D. (2014). The challenges of lean manufacturing implementation in SMEs. *Benchmarking: An International Journal, 21*(6), 987–1002.

Saaty, T. (2008). Decision making with analytic hierarchy process. *International Journal of Services Science, 1*(1), 83–98.

Sánchez, M. A., & Pérez, M. P. (2001). Lean indicators and manufacturing strategies. *International Journal of Operations & Production Management, 21*(11), 1433–1452.

Schröders, T., & Cruz-Machado, V. (2015). Sustainable lean implementation: An assessment tool. *Advances in Intelligent Systems and Computing, 362,* 1249–1264.

Searcy, D. (2009). Developing a lean performance score. *Strategic Finance, 91*(3), 34–39.

Seyedhosseini, S. M., Taleghani, A. E., Baksha, A., & Partovi, S. (2011). Extracting leanness criteria by employing the concept of balanced scorecard. *Expert System with Applications, 38,* 10454–10461.

Sezen, B., Karakadilar, I., & Buyukozkan, G. (2012). Proposition of a model for measuring adherence to lean practices: Applied to Turkish automotive part suppliers. *International Journal of Production Research, 50*(14), 3878–3894.

Shah, R., & Ward, P. T. (2003). Lean manufacturing: context, practice bundles, and performance. *Journal of Operations Management, 21*(2), 129–149.

Shah, R., & Ward, P. T. (2007). Defining and developing measures of lean production. *Journal of Operations Management, 25*(4), 785–805.

Shetty, D., Ahad, A., & Cummings, R. (2010). Survey-based spreadsheet model on lean implementation. *International Journal of Lean Six Sigma, 1*(4), 310–334.

Singh, B., Garg, S. K., & Sharma, S. K. (2010). Development of index for measuring leanness: study of an Indian auto component industry. *Measuring Business Excellence, 14*(2), 46–53.

Soriano-Meier, H., & Forrester, P. L. (2002). A model for evaluating the degree of leanness of manufacturing firms. *Integrated Manufacturing Systems, 13*(2), 104–109.

Srinivasaraghavan, J., & Allada, V. (2006). Application of mahalanobis distance as a lean assessment metric. *International Journal of Advanced Manufacturing Technology, 29,* 1159–1168.

Sterman, J. D. (2000). *Business dynamics systems thinking and modeling for a complex world.* New York, NY: McGraw-Hill.

Stone, K. B. (2010). Relationships between organizational performance and change factors and manufacturing firms' leanness. Ph.D. Dissertation, Colorado State University, Colorado, USA.

Stone, K. B. (2012). Lean transformation: Organizational performance factors that influence firms' leanness. *Journal of Enterprise Transformation, 2*(4), 229–249.

Strategos. (2010). Strategos Lean Assessment Tool. Available http://www.strategosinc.com/assessment.htm. Accessed June 2015.

Susilawati, A., Tan, J., Bell, D., & Sarwar, M. (2013). Develop a framework of performance measurement and improvement system for lean manufacturing activity. *International Journal of Lean Thinking, 4*(1), 51–64.

Susilawati, A., Tan, J., Bell, D., & Sarwar, M. (2015). Fuzzy logic based method to measure degree of lean activity in manufacturing industry. *Journal of Manufacturing Systems, 34,* 1–11.

Taj, S. (2005). Applying lean assessment tools in Chinese hi-tech industries. *Management Decision, 43*(4), 628–643.

Taleghani, A. E., Hosseini, S. M., & Bakhsha, A. (2010). Performance measurement of home appliances manufacturing company by leanness concept and system dynamics approach. *Business Research Yearbook: Global Business Perspectives, 18*(2), 640–647.

Toloie-Eshlaghy, A., & Kalantary, M. (2011). A mathematical model to auditing leanness by competitive benchmarking in an Iranian automaker. *Production Management, 37,* 3755–3757.

Vimal, K. E., & Vinodh, S. (2012). Thirty criteria based leanness assessment using fuzzy logic approach. *The International Journal of Advanced Manufacturing Technology, 60*(9–12), 1185–1195.

Vimal, K. E., & Vinodh, S. (2013). Application of artificial neural network for fuzzy logic based leanness assessment. *Journal of Manufacturing Technology Management, 24*(2), 274–292.

Vinodh, S., Arvind, K. R., & Somanaathan, M. (2010). Application of value stream mapping in an Indian camshaft manufacturing organisation. *Journal of Manufacturing Technology Management, 21*(7), 888–900.

Vinodh, S., & Balaji, S. R. (2011). Fuzzy logic based leanness assessment and its decision support system. *International Journal of Production Research, 49*(13), 4027–4041.

Vinodh, S., & Chintha, S. (2011). Leanness assessment using multi-grade fuzzy approach. *International Journal of Production Research, 49*(2), 431–445.

Vinodh, S., & Joy, D. (2012). Structural equation modelling of lean manufacturing practices. *International Journal of Production Research, 50*(6), 1598–1607.

Vinodh, S., Prakash, N., & Selvan, K. (2011). Evaluation of leanness using fuzzy association rules mining. *International Journal of Advanced Manufacturing Technology, 57,* 343–352.

Vinodh, S., & Vimal, K. (2012). Thirty criteria based leanness assessment using fuzzy logic approach. *International Journal of Advanced Manufacturing Technology, 60,* 1185–1195.

Wagner, C. L., Felipe, A. C., Mauro, L. J., & Robisom, D. C. (2014). Performance evaluation of lean manufacturing implementation in Brazil. *International Journal of Productivity and Performance Management, 63*(5), 529–549.

Wahab, A. N. A., Mukhtar, M., & Sulaiman, R. (2013). A conceptual model of lean manufacturing dimensions. *Procedia Technology, 11,* 1292–1298.

Wan, H. (2006). *Measuring leanness of manufacturing systems and identifying leanness target by considering agility.* Dissertation submitted to the faculty of the Virginia Polytechnic Institute and State University, USA.

Wan, H., & Chen, F. F. (2008). A leanness measure of manufacturing systems for quantifying impacts of lean initiatives. *International Journal of Production Research, 46*(23), 6567–6584.

Wan, H., & Chen, F. (2009). Decision support for lean practitioners: A web-based adaptive assessment approach. *Computers in Industry, 60,* 277–283.

Wan, H., Chen, F. F., & Rivera, L. (2007). Leanness score of value stream maps. In *Proceedings of the 2007 Industrial Engineering Research Conference Proceedings in Tennessee, USA, 19–23 May 2007.*

Wong, W. P., Ignatius, J., & Soh, K. L. (2014). What is the leanness level of your organisation in lean transformation implementation? An integrated lean index using ANP approach. *Production Planning and Control: The Management of Operations, 25*(4), 273–287.

Zahn, E., Dillerup, R., & Schmid, U. (1998). Strategic evaluation of flexible assembly systems on the basis of hard and soft decision criteria. *System Dynamics Review, 14*(4), 263–284.

Zanjirchi, S. M., Tooranlo, H. S., & Nejad, L. Z. (2010). Measuring organizational leanness using fuzzy approach. In *Proceedings of the 2010 International Conference on Industrial Engineering and Operations Management, in Dhaka, Bangladesh, 9–10 January 2010.*

Lean Supply Chain Management: A Systematic Literature Review of Practices, Barriers and Contextual Factors Inherent to Its Implementation

Satie Ledoux Takeda Berger, Guilherme Luz Tortorella and Carlos Manuel Taboada Rodriguez

Abstract The objective of this chapter is to perform a systematic literature review to identify the main Lean Supply Chain Management (LSCM) practices, barriers to such implementation and contextual factors that influence it. Through this literature examination, it is expected to identify the main gaps related to LSCM implementation and discuss the relevance of research in this topic, indicating future research directions. A systematic literature review (SLR) was devised and adopted, which involved the selection, classification, and evaluation of the literature, resulting in a final portfolio of 60 research articles. It is worth noticing that no temporal delimitation of publications was defined. The content of extant LSCM literature was critically analyzed and synthesized from the perspective of the practices, barriers and contextual factors inherent to LSCM implementation. The analysis of extant literature shows that there is a significant increase in studies related to LSCM, especially after 2011. Based on an extensive systematic review of the literature, we consolidated 18 practices, 12 barriers and 8 contextual factors inherent to LSCM implementation. It is noteworthy that most of the studies published to date on LSCM have focused on outlining practices and their potential benefits, inferring that once companies adopt them the lean implementation would be automatically started. However, such implementation throughout the supply chain is extremely difficult and challenging. From the analysis of this portfolio, it was also verified that most researchers address the three topics (practices, barriers and contextual factors) in an isolated way, not correlating them from a holistic perspective. This research expands previous work on LSCM, strengthening of the body of knowledge on the subject and consolidating the main practices of LSCM, barriers and contextual factors inherent to its implementation. The clear identification of these topics may

S. L. T. Berger (✉) · G. L. Tortorella · C. M. T. Rodriguez
Department of Industrial Engineering and Systems, Universidade Federal de Santa Catarina, Florianópolis, Brazil
e-mail: satietakeda@hotmail.com

G. L. Tortorella
e-mail: gtortorella@bol.com.br

C. M. T. Rodriguez
e-mail: tabcarlos@gmail.com

© Springer International Publishing AG, part of Springer Nature 2018   39
J. P. Davim (ed.), *Progress in Lean Manufacturing*,
Management and Industrial Engineering,
https://doi.org/10.1007/978-3-319-73648-8_2

help researchers and practitioners to anticipate occasional difficulties and set the proper expectations along the LSCM implementation.

Keywords Lean supply chain management · Practices · Barriers
Contextual factors · Systematic literature review

1 Introduction

The supply chain comprises all activities related to the flow and transformation of products and information, starting from raw materials to the end user, both downstream and upstream in the supply chain (Ballou 2009). According to Ugochukwu et al. (2012) and Christopher and Towill (2001), an appropriate Supply Chain Management (SCM) is key for companies, impacting their operational performance in terms of lower inventory level, higher customer satisfaction and processes efficiency, higher quality, reduced costs and improvements in delivery service level. Alves Filho et al. (2004) emphasize the increasing amount of studies that investigate the different contexts and practices related to SCM. Jasti and Kodali (2015a) emphasize that to assure an organization's competitiveness it is necessary to produce the right products, with the expected quality and quantity, at the right price and time, for the right customer. Within this scenario, due to the benefits provided to manufacturing environments, the incorporation of lean principles and practices into SCM has culminated in differentiated results along the supply chain, surpassing those already achieved by the organizations individually (Arif-Uz-Zaman and Ahsan 2014).

In this sense, the extension of the application of lean principles and practices to supply chain is called Lean Supply Chain Management (LSCM) (Anand and Kodali 2008). Vitasek et al. (2005) define LSCM as being a set of organizations directly linked by upstream and downstream flows of products, services, finances and information that work collaboratively aiming at the reduction of costs and waste, demonstrating in an efficient way what is necessary for customer's individual needs. According to Shah and Ward (2003), LSCM emphasizes the use of lean practices in a synergistic way to create a high-quality production and logistics systems that produce and deliver according to the customers demand and with little or no waste. Many studies are found addressing LSCM (e.g., Levy 1997; Sridharan et al. 2005; Li et al. 2006; Taylor 2006; Boonsthonsatit and Jungthawan 2015); but most of the applications are restricted to certain industrial segments, or approach only one lean practice, neglecting a holistic perspective inherent to the implementation of LSCM.

In this context, Jasti and Kodali (2015a) comment on the lack of a stable and unidirectional theory regarding LSCM concepts, since many studies focus only on individual aspects of LSCM, and few have a focus on both upstream and downstream activities of the organization system. Further, according to Anand and Kodali (2008), several modifications must be made to adapt lean principles and practices to SCM. While manufacturing predominantly involves the flow of materials with a reduced amount of information within the boundaries of the organization, the supply chain comprises the flow of materials, information, and resources beyond the boundaries of the organization. Thus, both the benefits and the barriers faced for LSCM implementation may differ from those already known in manufacturing (Manzouri and Rahman 2013). Additionally, the contextual factors that may influence the implementation of LSCM, such as sector and size of the supply chain, tier level, among others, are scarcely evidenced in the literature if compared to studies in manufacturing environments (Li et al. 2006). Overall, it is verified the incipience of the literature with respect to the addressed topic and, consequently, three research questions can be raised: (i) "what are the main practices of LSCM?"; (ii) "what are the inherent barriers to its implementation?"; and (iii) "what are the contextual factors that influence the LSCM implementation?".

Thus, the objective of this study is to perform a systematic literature review to identify the main LSCM practices, barriers to such implementation and contextual factors that influence it. Through this literature examination, it is expected to identify the main gaps related to LSCM implementation and discuss the relevance of research in this topic, indicating future research directions. Literature review is a widely used method to analyze comprehensively different approaches to the topic under study, as well as to reinforce the proposed research problem and to justify the differential of the proposal by reorganizing existing knowledge and identifying gaps (Paré et al. 2015). The literature review presented in this article applies explicit and systematized search methods, synthesis of the selected information, and integrates the information of a set of studies carried out separately. Besides this introduction section, this chapter is structured as follows: Sect. 2 briefly conceptualizes fundamental principles that underlie SCM; in Sect. 3, the proposed literature research method is described, whose results are discussed in Sect. 4. Finally, Sect. 5 presents the final remarks and directions for future research.

2 Literature Review

The term "Supply Chain Management" was first proposed in the literature in the 1980s, but it was only in the 1990s that the first organizational reports were actually evidenced. SCM implies a management change from exclusive improvement efforts oriented to internal problems, to focus on the relations with the other companies that are part of the organization's supply chain (Alves Filho et al. 2004). For the Council of Supply Chain Management Professionals (2013), "SCM encompasses the planning and management of all activities involved in supply and acquisition, conversion and all logistics management activities. It also includes coordination and collaboration with channel partners, who can be suppliers, intermediaries, service providers and customers." Further, SCM can be seen as a way to efficiently connect each agent of the manufacturing and supply processes, from the raw material to the final consumer. Sridharan et al. (2005) comment that SCM aims to integrate the various structures and processes, facilitating and coordinating the flow of goods, services, and information necessary to provide the value that customers want. Complementarily, SCM focuses on how companies manage their technology, information, and skills to improve their competitive advantage (Ariffin et al. 2015).

Akkermans et al. (2004) comment that SCM is a high-complex activity from both academic' and practitioner's perspective. This is justified by the fact that a supply chain is composed by a net of companies or independent business units, starting from the original supplier to the final customers, whose management becomes a broad and challenging task (Lambert et al. 2005; Ellram and Cooper 2014). Thus, to accomplish an efficient SCM, it is important to know and understand how organizations are structured. In this sense, two fundamental aspects of the supply chain structure are suggested (Lambert et al. 1998; Lambert and Cooper 2000): supply chain agents and their structural dimensions. The agents of a supply chain include all organizations with which the focal company interacts directly or indirectly through its suppliers or customers, from the raw material acquisition to the final consumer (Lambert and Cooper 2000). Regarding the structural dimensions, Lambert et al. (1998) emphasize three dimensions as essential: (i) horizontal structure, which refers to the number of tiers in the entire supply chain; (ii) vertical structure, denoting the number of suppliers and customers belonging to the same tier; and (iii) horizontal position, which is the position of the focal company within the supply chain.

3   Method

The adopted method is comprised of 6 steps, as shown in Table 1.

In the first step, the research question, which the paper will seek to answer, was defined and served as the starting point for the systematic review of the literature. This study consists of three questions already presented in the introductory section. Then, in the second step, the databases to be used in the search were determined. The bases were chosen according to their availability in CAPES (Brazilian Coordination for Improvement of Higher Level Personnel) Journals website.

Table 1 General description of the systematic literature review process

Steps	Description for conducting a structured literature review
1	Define the research question
2	Choose the databases to be consulted; set keywords and search strategies
3	Define criteria for the selection of articles and conduct search in selected databases
4	Define the initial portfolio of articles, applying the criteria in the selection of articles and get the final portfolio of articles
5	Critically analyze all studies in the review and describe a critical view by the selected articles
6	Determinate a final considerations and contributions of this research

The strategies defined for the selection of databases were delimited according to Lancaster (2004), who argues that a bibliographic database should be evaluated for its usefulness in responding to the following information needs: (i) coverage, how complete is the content of the database in relation to a subject; (ii) recoverability, how many documents on the subject can be found in the database using a search strategy that is not very complex; (iii) predictability, the researcher can verify with efficiency the relevance of the documents from the information contained in the database; and (iv) actuality, the database has a good frequency in the inclusion of new publications. In this sense, three bases were defined: Emerald, Scopus, and Web of Science. Further, different combinations of keywords were established for the initial search, as presented in Table 2. It is worth noticing that no temporal delimitation of publications was defined.

In the third step, some criteria were established for the selection of the articles from the initial portfolio, namely: (i) eliminate duplicates or nonscientific articles; (ii) verify alignment of articles' titles with the research theme; (iii) check alignment of abstracts to the research theme; and (iv) assure that articles' full texts are aligned with the research theme. Then, based on the selected keywords, a search was performed on the chosen databases according to the previously defined strategies. In step four, the analysis of the initial portfolio was performed, which totalized 1384 references. In addition, these references have been exported to the Endnote X7® management tool to assist with their organization. Subsequently, the four criteria defined in step three were applied in the selection of articles, in order to exclude those that were not aligned with the research objectives. Hence, 867 duplicate or nonscientific articles were excluded, as well as 457 articles unrelated to

Table 2 Bibliographic research and selection of articles

Keywords	Quantitative databases		
	Emerald	Scopus	Web of science
"Lean supply" AND *"practices"* OR *"implement"*	234	58	25
"Lean supply" AND *"failures"* OR *"challenges"* OR *"barriers"*	175	40	8
"Lean supply" AND *"contextual factors"*	189	8	18
"Lean supply"	282	236	111
Total at each base	**880**	**342**	**162**
Total (Initial Portfolio)	**1384**		
Duplicate or non-scientific articles	867		
Articles not related to the objectives of the work (title and abstract)	401		
Articles not related to the objectives of the work (full text)	56		
Total selected articles (Final Portfolio)	**60**		

the topic, through the analysis of titles, abstracts or full text. In total, 1323 articles were removed, remaining only 60 as the final portfolio. In step five, a critical reading and evaluation of the final defined articles portfolio were performed; and a critical summary was developed in order to highlight the relevant findings available in these articles. Finally, step six allowed the determination of the final considerations and contributions of this study, and the establishment of future research directions on the theme.

4 Discussion on LSCM Research

4.1 Overall Analysis of the Final Portfolio

The studies consolidated in the final portfolio of LSCM articles indicate research in different supply chains, such as food (Vlachos 2015), toys (Yew Wong et al. 2005), electronics (McIvor 2001), automotive (Adamides et al. 2008; Wee and Wu 2009; Boonsthonsatit and Jungthawan 2015), agribusiness (Taylor 2006; Perez et al. 2010), among others. However, according to Cudney and Elrod (2011), some supply chains still have a shortage of studies, such as informatics, civil construction, design, engineering, government, and military.

As for the temporal aspect, Ugochukwu et al. (2012) comment that, although the lean principles and practices became popular in 1990s, their implementation in the supply chain context gained more attention a few years later. According to the authors, the extension of lean principles and practices to supply chain can be attributed to the publication of the book "Lean Thinking" by Womack and Jones (1996) in which it promotes lean implementation throughout the supply chain. Hines et al. (2004) indicate that the understanding of lean principles and practices has undergone an evolution over the years, starting from the approach focused on plant floor tools to a contingent perspective along the value chain. This fact is observed in the final portfolio, since the first article related to the subject was published in 1996 by Lamming, with a discrete increase of publications in the following years. However, after 2011 it has been noticed a significant increase in the number of published articles related to this subject, with the highest number of publications (8 articles) on 2015.

Regarding the research method among the articles included in the final portfolio, applied research prevailed (70%) followed by theoretical research (30%). According to Vilaça (2010), applied research is usually focused on solving practical issues. On the other hand, the theoretical research is essentially based on a bibliographical research, which provides analysis and discussion of a predefined theme. Further, among the 42 applied research articles found, 15 correspond to studies in emerging economies, while 14 are studies in developed countries.

4.2 LSCM Practices

Lean practices can be applied across the entire supply chain, from placing the order with suppliers to distributing and delivering the product to the end customer. Previous studies (Wee and Wu 2009; Perez et al. 2010; Manzouri 2012; Boonsthonsatit and Jungthawan 2015; Hartono et al. 2015; Vlachos 2015) associate the implementation of LSCM practices with improvements in the supply chain's operational performance, regardless of its context. Erridge and Murray (1998), for instance, indicate that through the application of LSCM practices similar benefits to those of the manufacturing industry can be obtained in the Belfast City Hall, the main city of Northern Ireland. These benefits can be observed in terms of reduction of inventory, increase in services quality, cost reduction and better relationship with suppliers and customers. However, studies on LSCM practices are still less frequently evidenced in the literature if compared to manufacturing environments. The implementation of LSCM practices is considered more complex than manufacturing (Martínez-Jurado and Moyano-Fuentes 2014), since they require a significant adaptation and involve different companies (Anand and Kodali 2008; Manzouri et al. 2014). In this sense, most of the studies that address LSCM practices indicate the need for leadership restructuring and establishment of a supporting infrastructure (Yew Wong et al. 2005; Adamides et al. 2008; Behrouzi and Wong 2011; Vlachos 2015). It is worth noticing that there are some industry sectors little explored regarding the progress of LSCM implementation, in which different challenges and benefits may emerge than those already expected (Cudney and Elrod 2011).

Further, some studies (e.g., Taylor 2006; Anand and Kodali 2008; Vlachos 2015) intend to connect LSCM practices and lean principles. Perez et al. (2010), for example, evaluate the relationship of contextual variables and performance of a supply chain with LSCM practices. The authors propose a structure with seven dimensions of LSCM practices categorized into five lean principles: (i) definition of value, (ii) identification of the value stream, (iii) making the value flow, (iv) pulling the value from the customer's demand and (v) seek perfection. The seven proposed dimensions are: (i) demand management; (ii) specification of the value; (iii) standardization of processes and products; (iv) value chain efficiency; (v) key process indicators; (vi) establishing alliances with members of the chain; and (vii) cultural change.

Another important aspect concerns the lack of holistic approaches to the implementation of LSCM practices. Many studies deal only with individual aspects of LSCM, presenting a narrow or isolated perspective of activities upstream and downstream of the flow (Martínez-Jurado and Moyano-Fuentes 2014). Among the final portfolio, only two studies proposed broad conceptual frameworks regarding the implementation of LSCM practices. Anand and Kodali (2008), later complemented by Jasti and Kodali (2015a), suggest eight pillars for the implementation of LSCM, which are constituted by 82 practices; they are: (i) management of information technology; (ii) management of suppliers; (iii) waste disposal;

(iv) JIT production; (v) customer relationship management; (vi) logistics management; (vii) commitment of senior management; (viii) continuous improvement. However, the proposed conceptual framework was not empirically validated, characterizing a research gap.

Overall, Tables 3, 4 and 5, consolidate the most cited LSCM practices (P) according to 55 articles from the final portfolio. The citation frequency of these 18 practices presents significant variations. Practices P1 (Kanban or pull system) and P2 (Close relationship between customer, supplier and relevant stakeholders) appear to be the most frequently cited in the LSCM studies, with 38 and 32 citations, respectively. The high frequency of citations can be explained by the fact that these practices are included in the precursor studies of LSCM (e.g., Lamming 1996; Erridge and Murray 1998), since their impact on both manufacturing process and supply chain performance can be more easily perceived. In fact, these practices were consistently associated with the LSCM studies over time, leading to high number of research evidence that reports their application. Specifically for P1, McIvor (2001) comments on its impact on obtaining lower inventory levels and greater visibility of quality problems. In addition, such practice is commonly associated with just-in-time (JIT) deliveries (Dües et al. 2013; Wiengarten et al. 2013), in which the right material is delivered at the expected time, place and quantity (Qrunfleh and Tarafdar 2013). Consequently, the adoption of P1 implies a narrowing of information and material flows between suppliers and customers, reinforcing the collaboration between them (P2) (Martínez-Jurado and Moyano-Fuentes 2014). In this sense, it is reasonable to expect that both P1 and P2 present high recurrence of citations over time, since they are closely related (Bhamu and Singh Sangwan 2014).

On the other hand, P18 (Establishment of distribution centers) showed to be the least mentioned in the literature, presenting only 3 citations out of 55 studies. The implementation of distribution centers is generally motivated due to potential impacts on transportation costs and order processing (Baker 2004). Although the first studies on this practice date from the 1970s (La Londe et al. 1971), their association and later insertion in the LSCM approach are relatively more recent. In fact, Taylor (2006) appears to be the first study to suggest the incorporation this practice into the set of LSCM practices. However, only in Sharma et al. (2015) and Jasti and Kodali (2015a) that P18 was explicitly included in the set of LSCM practices. Thus, from the increased understanding and expansion of lean thinking to supply chains, which provided a much more comprehensive approach to LSCM implementation, P18 gained considerable attention and began to be treated as a LSCM practice.

In general, the 18 practices consolidated in Tables 3, 4 and 5 emerge from an extensive review of the literature and provide a representative view of the main practices adopted in LSCM. The approach of analyzing the impact of lean implementation based on the assessment of the adoption level of predefined practices has been widely used in previous studies (Qi and Chu 2009; Rahman et al. 2010; Manzouri et al. 2013; Sharma et al. 2015) and seems to be also quite effective in understanding companies maturity regarding LSCM.

Table 3 LSCM practices and their frequency of citation in the literature (Part I)

	Practices/Authors	(1)	(2)	(3)	(4)	(5)	(6)	(7)	(8)	(9)	(10)	(11)	(12)	(13)	(14)	(15)	(16)	(17)	(18)	(19)	(20)
P_1	Kanban or pull system	x	x		x			x		x			x	x	x	x		x	x	x	x
P_2	Close relationship between customer, supplier and relevant stakeholders	x	x		x	x	x			x		x	x		x	x	x			x	x
P_3	Use of information technology to share and integrate the flow of information along the supply chain (e.g., EDI, RFID, ERP, etc.)		x		x		x	x	x	x			x	x	x		x		x		x
P_4	Efficient and continuous replenishment	x						x		x				x				x	x	x	x
P_5	Continuous improvement		x			x							x		x	x		x			x
P_6	Value chain analysis or value stream mapping									x	x			x	x			x			
P_7	Keiretsu (relationship based on trust, cooperation and educational support, with suppliers playing an important strategic role in the organization)	x	x						x	x			x				x		x	x	
P_8	Supplier managed inventory (consigned)	x				x				x			x		x		x			x	

(continued)

Table 3 (continued)

Practices/Authors		(1)	(2)	(3)	(4)	(5)	(6)	(7)	(8)	(9)	(10)	(11)	(12)	(13)	(14)	(15)	(16)	(17)	(18)	(19)	(20)
P_9	Distribution logistics	x			x									x	x		x				
P_{10}	Standardized work procedures to assure quality achievement		x									x	x		x			x			x
P_{11}	Frequent participation from the beginning of the project and new product development process		x	x	x		x						x								x
P_{12}	Hoshin Kanri (development of strategies and commitment and support of the top managers)	x				x				x	x	x	x								x
P_{13}	Development of supply chain KPIs								x	x		x				x					x
P_{14}	Leveled scheduling or heijunka	x												x				x			
P_{15}	Kyoryoku Kai (association of suppliers working together to develop more efficient methods of work reducing waste)	x								x					x						x

(continued)

Table 3 (continued)

Practices/Authors	(1)	(2)	(3)	(4)	(5)	(6)	(7)	(8)	(9)	(10)	(11)	(12)	(13)	(14)	(15)	(16)	(17)	(18)	(19)	(20)
P_{16} Problems solution (frequent feedback working together for shared solutions)				x															x	
P_{17} Value chain management									x				x	x						x
P_{18} Establishment of distribution centers									x											

Authors (1) Lamming (1996), (2) Erridge and Murray (1998), (3) McIvor (2001), (4) Arkader (2001), (5) Huang et al. (2002), (6) Birgün Barla (2003), (7) Yew Wong et al. (2005), (8) Jaklic et al. (2006), (9) Taylor (2006), (10) Eisler et al. (2007), (11) Morgan (2007), (12) Machado and Pereira (2008), (13) Adamides et al. (2008), (14) Anand and Kodali (2008), (15) Found et al. (2008), (16) Qi and Chu (2009), (17) Wee and Wu (2009), (18) Parveen and Rao (2009), (19) Stavrulaki and Davis (2010), (20) Perez et al. (2010)

Table 4 LSCM practices and their frequency of citation in the literature (Part II)

Practices/Authors	(21)	(22)	(23)	(24)	(25)	(26)	(27)	(28)	(29)	(30)	(31)	(32)	(33)	(34)	(35)	(36)	(37)	(38)	(39)	(40)
P_1 *Kanban* or pull system	x	x		x	x		x	x	x	x		x	x	x	x		x	x	x	x
P_2 Close relationship between customer, supplier and relevant stakeholders		x	x			x		x	x			x	x	x	x	x	x	x		x
P_3 Use of information technology to share and integrate the flow of information along the supply chain (e.g., EDI, RFID, ERP, etc.)	x		x			x		x	x					x				x		x
P_4 Efficient and continuous replenishment	x	x			x			x	x		x			x	x		x	x		
P_5 Continuous improvement		x							x	x		x	x				x	x		x
P_6 Value chain analysis or value stream mapping		x		x			x		x	x					x		x	x	x	
P_7 *Keiretsu* (relationship based on trust, cooperation and educational support, with			x		x	x		x				x			x	x				

(continued)

Table 4 (continued)

Practices/Authors		(21)	(22)	(23)	(24)	(25)	(26)	(27)	(28)	(29)	(30)	(31)	(32)	(33)	(34)	(35)	(36)	(37)	(38)	(39)	(40)
	suppliers playing an important strategic role in the organization	x																			
P8	Supplier managed inventory (consigned)					x	x						x	x		x				x	x
P9	Distribution logistics			x			x	x		x			x								
P10	Standardized work procedures to assure quality achievement				x													x	x	x	x
P11	Frequent participation from the beginning of the project and new product development process			x			x							x			x				
P12	Hoshin Kanri (development of strategies and commitment and support of the top managers)																x	x			
P13	Development of supply chain KPIs					x					x				x						

(continued)

Table 4 (continued)

Practices/Authors		(21)	(22)	(23)	(24)	(25)	(26)	(27)	(28)	(29)	(30)	(31)	(32)	(33)	(34)	(35)	(36)	(37)	(38)	(39)	(40)
P_{14}	Leveled scheduling or *heijunka*				x													x	x		
P_{15}	*Kyoryoku Kai* (association of suppliers working together to develop more efficient methods of work reducing waste)								x												
P_{16}	Problems solution (frequent feedback working together for shared solutions)																x	x			
P_{17}	Value chain management																				
P_{18}	Establishment of distribution centers																				

Authors (21) Rahman et al. (2010), (22) Parveen et al. (2011), (23) Gueimonde-Canto et al. (2011), (24) Cudney and Elrod (2011), (25) Carvalho et al. (2011), (26) Manzouri (2012), (27) Al-Aomar (2012), (28) Azevedo et al. (2012), (29) Drohomeretski et al. (2012), (30) Karim and Arif-Uz-Zaman (2013), (31) Camacho-Miñano et al. (2013), (32) Manzouri and Rahman (2013), (33) Qrunfleh and Tarafdar (2013), (34) Wiengarten et al. (2013), (35) Dües et al. (2013), (36) Martinez-Jurado and Moyano-Fuentes (2014), (37) Hadid and Mansouri (2014), (38) Bhamu and Singh Sangvan (2014), (39) Arif-Uz-Zaman and Ahsan (2014), (40) Manzouri et al. (2014)

Table 5 LSCM practices and their frequency of citation in the literature (Part III)

	Practices/Authors	(41)	(42)	(43)	(44)	(45)	(46)	(47)	(48)	(49)	(50)	(51)	(52)	(53)	(54)	(55)	Frequency of citation
P_1	Kanban or pull system	x		x	x		x	x	x		x		x	x		x	38
P_2	Close relationship between customer, supplier and relevant stakeholders			x	x		x			x					x	x	32
P_3	Use of information technology to share and integrate the flow of information along the supply chain (e.g., EDI, RFID, ERP, etc.)			x	x	x			x						x	x	26
P_4	Efficient and continuous replenishment	x			x	x	x				x		x	x			25
P_5	Continuous improvement	x		x	x	x	x	x	x		x				x		24
P_6	Value chain analysis or value stream mapping	x	x	x	x	x	x	x	x		x	x			x		23
P_7	Keiretsu (relationship based on trust, cooperation and educational support, with suppliers playing an important strategic role in the organization)			x	x		x			x		x					21
P_8	Supplier managed inventory (consigned)			x	x		x		x						x		20
P_9	Distribution logistics			x	x	x	x		x			x			x		18
P_{10}	Standardized work procedures to assure quality achievement	x		x	x	x					x	x	x				17

(continued)

Table 5 (continued)

Practices/Authors	(41)	(42)	(43)	(44)	(45)	(46)	(47)	(48)	(49)	(50)	(51)	(52)	(53)	(54)	(55)	Frequency of citation
P_{11} Frequent participation from the beginning of the project and new product development process				x		x			x		x					13
P_{12} *Hoshin Kanri* (development of strategies and commitment and support of the top managers)				x							x					12
P_{13} Development of supply chain KPIs				x	x	x					x					12
P_{14} Leveled scheduling or *heijunka*				x	x	x									x	10
P_{15} *Kyoryoku Kai* (association of suppliers working together to develop more efficient methods of work reducing waste)				x				x					x			8
P_{16} Problems solution (frequent feedback working together for shared solutions)				x		x						x				7
P_{17} Value chain management				x	x		x									7
P_{18} Establishment of distribution centers				x		x										3

Authors (41) Jasti and Kodali (2015a), (42) Boonsthonsatit and Jungthawan (2015), (43) Hartono et al. (2015), (44) Jasti and Kodali (2015b), (45) Olesen et al. (2015), (46) Sharma et al. (2015), (47) Vlachos (2015), (48) Adebanjo et al. (2016), (49) Jajja et al. (2016), (50) Dora et al. (2016), (51) Soni and Kodali (2016), (52) Marodin et al. (2016), (53) Carvalho et al. (2017), (54) Duarte and Machado (2017), (55) Bevilacqua et al. (2017)

4.3 Barriers and Critical Factors for LSCM Implementation

A barrier is an obstacle that prevents or restricts progress, making it difficult to achieve something (Kumar et al. 2016). To be successful in any organizational change, the existing barriers need to be identified and understood (Jadhav et al. 2014). The LSCM implementation, like any other improvement initiative, entails enormous difficulties (Rahman et al. 2010). Although LSCM has been applied in different segments in the last decades, a few questions remain unanswered due to the inherent supply chain complexity and longer-term results, entailing additional challenges to improvements implementation throughout the chain (Adebanjo et al. 2016). On the other hand, the same studies allow distinguishing the particularities of the applied practices and also the barriers faced in LSCM adoption (Yew Wong et al. 2005).

From the 60 articles reviewed, only 34 addressed some kind of barrier inherent to the LSCM implementation. Out of these 34, only four articles explicitly presented the barriers and challenges of LSCM as the main theme (e.g., McIvor 2001; Manzouri et al. 2013; Jadhav et al. 2014; Dora et al. 2016). The remaining articles shallowly approached a few barriers resulting from the application of LSCM practices (e.g., Arkader 2001; Anand and Kodali 2008; Perez et al. 2010; Adebanjo et al. 2016). Vlachos (2015) describes the implementation of LSCM practices in a tea company in the United Kingdom, highlighting the difficulties encountered. The author reports the lack of involvement of top management in the improvement projects, implying a limited and failed implementation. Jadhav et al. (2014) comment that the only way to create a truly lean transformation is through a strong leadership at the top of the organization, including the CEO. Hence, the actual involvement of top managers is fundamental to support and sustain improvements (Yew Wong et al. 2005). In turn, lack of commitment may lead to a number of issues, such as limited access to resources, lengthy decision-making processes and communication failures (Perez et al. 2010).

Another important aspect to consider is the development of specialized teams, which are usually focused on developing training on lean principles and practices, empowering employees with the required knowledge and skills (Karim and Arif-Uz-Zaman 2013). As lean implementation becomes reasonably consolidated within the organization (shop floor and business processes), most companies extend training to agents of their supply chain (Cudney and Elrod 2011). However, the extension of LSCM implementation tends to be initially focused on upstream agents (suppliers) and their practices (Bevilacqua et al. 2017). Thus, the existence of specialized teams for training and qualification on LSCM practices allows greater proximity with supply chain agents, establishing a development process that goes beyond the traditional issues related to price and delivery (Dües et al. 2013; Wiengarten et al. 2013; Martínez-Jurado and Moyano-Fuentes 2014). In this sense, some studies report that it is not possible to successfully implement LSCM practices without directly involving these agents (suppliers and customers) (McIvor 2001; Taylor 2006; Jajja et al. 2016).

In addition, several benefits have been indicated from the relationship enhancement of suppliers and customers. Consequently, the absence or lack of emphasis on these relationships can lead to a significant barrier to LSCM implementation (Vlachos 2015). According to Qrunfleh and Tarafdar (2013), it is imperative that managers develop strategic relationships based on trust with suppliers and customers. Distrust and hostility among these agents may discourage efforts to implement LSCM practices, implying failure of continuous improvement (Taylor 2006). Further, Manzouri et al. (2013) identified that the lack of trust among supply chain agents is an important barrier to overcome, as it undermines the information sharing process.

Taylor (2006) argues that there is a difficulty in moving away from current negotiation strategies, which is characterized by seeking profit maximization in the short term. Such strategies negatively influence the establishment of long-term partnerships and reinforce power-based relationships that jeopardize LSCM implementation (Perez et al. 2010). Although there are different levels of bargaining power among supply chain agents, gaining advantage over others is not coherent for an LSCM implementation, since it harms the development of shared benefits (Lamming 1996). Thus, enhancing trustful relationships among these agents also mitigates the risks to all parties. Therefore, it is reasonable to assume that these agents depend on each other to obtain higher levels of operational performance (Manzouri et al. 2014).

Tables 6 and 7 display the underlying barriers (B) to LSCM implementation synthesized from 34 articles included the final portfolio. It is worth noticing that there is a variable citation frequency for each barrier. Barrier B1 (Difficulties for cultural change) appears to be the most frequently mentioned during the LSCM implementation. In general, a successful lean implementation is highly dependent on the sociocultural aspects of an organization (Jadhav et al. 2014). Changes in an organizational culture represent shifts in norms and collective behaviors that encompass trust, hierarchy, work environment, communication, and fellowship (Dora et al. 2016). Cultural change is one of the greatest challenges for sustaining lean practices, whether in the organization or in the supply chain, and consequently, employees involvement is extremely important, as they are considered to be a valuable source of improvement ideas (Perez et al. 2010; Behrouzi and Wong 2011). A persistent obstacle that hinders the successful application of LSCM practices is the resistance of those who are asked to adopt their practices and principles (Perez et al. 2010). Complementarily (Lamming 1996) suggests that LSCM implementation requires a change in existing culture, introducing new cultural visions of collaboration and human resource management, while supports the achievement of long and short-term goals through the encouragement of employees (Dora et al. 2016; Kumar et al. 2016). Therefore, the high frequency of citation associated to B1 (76%) in the final portfolio is justified.

In opposition, the least cited barrier was B12 (Low understanding of concepts and principles related to LSCM implementation), with only 9% of citations. A successful LSCM implementation presupposes proper understanding of its principles and practices (Manzouri 2012; Manzouri et al. 2013). Anand and Kodali (2008)

Table 6 LSCM barriers and their frequency of citation in the literature (Part I)

Barriers/Authors		(1)	(2)	(3)	(4)	(5)	(6)	(7)	(8)	(9)	(10)	(11)	(12)	(13)	(14)	(15)	(16)	(17)	(18)	(19)	(20)
B_1	Difficulties for cultural changes	x	x	x		x	x		x		x	x		x	x		x	x	x		x
B_2	Lack of commitment of senior management					x	x		x	x	x	x	x	x			x	x	x	x	
B_3	Lack of specialized team development								x		x		x				x	x	x	x	
B_4	Lack of trust in supply chain partnerships	x		x	x	x		x	x				x				x		x		
B_5	High oscillation of demand					x	x	x	x		x				x	x	x				
B_6	Low information sharing			x		x		x	x									x			
B_7	Lack of collaboration and involvement of the entire supply chain				x		x	x	x	x			x	x							x
B_8	Lack of availability of resources				x				x											x	
B_9	Lack of Information and Communication Technology (ICT) infrastructure for integration												x	x					x		
B_{10}	Resistance to joining long-term strategies			x											x	x	x	x	x		

(continued)

Table 6 (continued)

Barriers/Authors		(1)	(2)	(3)	(4)	(5)	(6)	(7)	(8)	(9)	(10)	(11)	(12)	(13)	(14)	(15)	(16)	(17)	(18)	(19)	(20)
B_{11}	Complexity of the supply chain	x						x				x			x	x					
B_{12}	Low understanding of concepts and principles related to LSCM implementation													x						x	

Authors (1) Lamming (1996), (2) Erridge and Murray (1998), (3) McIvor (2001), (4) Arkader (2001), (5) Huang et al. (2002), (6) Yew Wong et al. (2005), (7) Jaklic et al. (2006), (8) Taylor (2006), (9) Eisler et al. (2007), (10) Morgan (2007), (11) Machado and Pereira (2008), (12) Adamides et al. (2008), (13) Anand and Kodali (2008), (14) Wee and Wu (2009), (15) Stavrulaki and Davis (2010), (16) Perez et al. (2010), (17) Cudney and Elrod (2011), (18) Behrouzi and Wong (2011), (19) Manzouri (2012), (20) Azevedo et al. (2012)

Table 7 LSCM barriers and their frequency of citation in the literature (Part II)

	Barriers/Authors	(21)	(22)	(23)	(24)	(25)	(26)	(27)	(28)	(29)	(30)	(31)	(32)	(33)	(34)	Frequency of citation
B_1	Difficulties for cultural changes	x	x	x	x	x	x		x		x	x	x	x	x	26
B_2	Lack of commitment of senior management		x	x	x		x	x		x	x	x	x	x		22
B_3	Lack of specialized team development	x	x	x	x	x	x	x	x	x	x	x	x	x	x	21
B_4	Lack of trust in supply chain partnerships		x		x		x	x				x				14
B_5	High oscillation of demand	x					x									11
B_6	Low information sharing			x	x			x			x		x		x	11
B_7	Lack of collaboration and involvement of the entire supply chain			x	x											10
B_8	Lack of availability of resources	x	x		x			x			x		x		x	10
B_9	Lack of Information and Communication Technology (ICT) infrastructure for integration			x				x			x		x	x		10
B_{10}	Resistance to joining long-term strategies															6
B_{11}	Complexity of the supply chain		x													6
B_{12}	Low understanding of concepts and principles related to LSCM implementation		x													3

Authors (21) Karim and Arif-Uz-Zaman (2013), (22) Manzouri et al. (2013), (23) Martinez-Jurado and Moyano-Fuentes (2014), (24) Jadhav et al. (2014), (25) Hadid and Mansouri (2014), (26) Bhamu and Singh Sangvan (2014), (27) Manzouri et al. (2014), (28) Tortorella et al. (2015), (29) Vlachos (2015), (30) Adebanjo et al. (2016), (31) Jajja et al. (2016), (32) Dora et al. (2016), (33) Kumar et al. (2016), (34) Bevilacqua et al. (2017)

argue that concepts related to LSCM are still not fully developed, especially in terms of its theoretical basis and elements, and the ways of its implementation. However, due to the lack of studies that deepen such aspects, the level of understanding and awareness regarding LSCM implementation is still very shallow (Manzouri et al. 2013), justifying the low citation of this barrier.

In general, the 12 barriers encountered emerge from the extensive literature review and provide a representative view of the main barriers inherent to LSCM implementation. Thus, the identification of these barriers can be a starting point to properly addressing the difficulties, allowing the anticipation of a few of them (Jadhav et al. 2014).

4.4 Contextual Factors for LSCM Implementation

Contextual factors are aspects or elements that influence the performance of a management system, which is conditioned by the specific characteristics of a company or its environment, such as number of employees, sales volumes, sector, time in which a management system is implemented and so on (Hadid and Mansouri 2014). Further, these factors represent situational characteristics usually exogenous to the focal organization or manager (Tortorella et al. 2015). A number of contextual factors are inherent to each supply chain and may affect the relationship between the cooperation level of its agents and their performance (Gueimonde-Canto et al. 2011). However, the modification of these factors tends to be limited and only possible with a long-term effort (Manzouri et al. 2013). Thus, taking into account their influence is vital for a better understanding of the LSCM implementation (Camacho-Miñano et al. 2013).

According to Karim and Arif-Uz-Zaman (2013), the proper selection of LSCM practices depends on the context of each company and its supply chain. Therefore, the strategy for the transition from a traditional supply chain model to a LSCM cannot be indiscriminately generalized, since the different contextual factors are determinant for such decision (Rahman et al. 2010). In this sense, Tables 8 and 9 compiles the contextual factors (CF) inherent to LSCM implementation. From the 60 articles included in the final portfolio, only 30 explicitly addressed some CF. It is worth noticing that, besides the variable citation frequency of each CF, the total citation frequency and the number of articles that addressed the subject was significantly lower than those that approached LSCM practices and barriers.

The contextual factor CF1 (Company size) was the most cited factor, as 63% of the articles indicated. Hadid and Mansouri (2014) comment that larger companies are more likely to have higher adoption levels of LSCM practices, since they usually have a more complex supply chain and, hence, need a more efficient management. Manzouri (2012) also claims that companies' size is positively associated with LSCM implementation, since larger organizations presuppose higher bargain power and leadership within the supply chain they belong. In turn, the contextual factor CF8 (Production volume) was the least cited among the studied authors, with only

Table 8 Contextual factors of LSCM and its citation frequency in the literature (Part I)

	Contextual factors/Authors	(1)	(2)	(3)	(4)	(5)	(6)	(7)	(8)	(9)	(10)	(11)	(12)	(13)	(14)	(15)
CF_1	Company size						x		x	x	x	x		x	x	
CF_2	Trained multifunctional team		x	x	x		x				x	x	x	x		
CF_3	Geographic location	x		x		x		x					x	x		x
CF_4	Supply chain sector								x				x	x		x
CF_5	Country's socio-economic factors	x		x				x								x
CF_6	Educational level	x	x								x					
CF_7	Plant age															
CF_8	Production volume															

Authors (1) Arkader (2001), (2) Huang et al. (2002), (3) Taylor (2006), (4) Machado and Pereira (2008), (5) Adamides et al. (2008), (6) Anand and Kodali (2008), (7) Found et al. (2008), (8) Qi and Chu (2009), (9) Wee and Wu (2009), (10) Perez et al. (2010), (11) Rahman et al. (2010), (12) Gueimonde-Canto et al. (2011), (13) Cudney and Elrod (2011), (14) Manzouri (2012), (15) Azevedo et al. (2012)

Table 9 Contextual factors of LSCM and its citation frequency in the literature (Part II)

Contextual factors/ Authors		(16)	(17)	(18)	(19)	(20)	(21)	(22)	(23)	(24)	(25)	(26)	(27)	(28)	(29)	(30)	Frequency of citation
CF_1	Company size	x	x	x			x	x	x	x		x	x	x		x	19
CF_2	Trained multifunctional team	x						x		x			x		x		11
CF_3	Geographic location					x					x	x					10
CF_4	Supply chain sector		x	x							x		x				8
CF_5	Country's socio-economic factors		x	x								x		x			8
CF_6	Educational level	x			x					x							6
CF_7	Plant age						x			x		x					3
CF_8	Production volume	x											x				2

Authors (16) Karim and Arif-Uz-Zaman (2013), (17) Camacho-Miñano et al. (2013), (18) Manzouri et al. (2013), (19) Martínez-Jurado and Moyano-Fuentes (2014), (20) Jadhav et al. (2014), (21) Hadid and Afshin Mansouri (2014), (22) Bhamu and Singh Sangwan (2014), (23) Manzouri et al. (2014), (24) Tortorella et al. (2015), (25) Adebanjo et al. (2016), (26) Jajja et al. (2016), (27) Dora et al. (2016), (28) Marodin et al. (2016), (29) Duarte and Machado (2017), (30) Bevilacqua et al. (2017)

7% of citations. High production volumes generally imply a greater prominence of the company within its supply chain. Such importance can affect the way in which the relationships between the company and its customers and suppliers are established, given the greater implicit bargaining power (Lamming 1996). However, studies that address the influence of production volume on LSCM implementation are still scarce (Karim and Arif-Uz-Zaman 2013), which indicates opportunities for future research development. Further, Dora et al. (2016) comment that many studies adopt a fragmented approach to LSCM implementation, ignoring their systemic nature and resulting in their failure. Thus, the 8 CF found from the literature review provide a representative view of the LSCM implementation. The identification of these CF, as well as their effect on LSCM implementation, is essential for the adaptation and customization of the improvement strategies adopted. Therefore, the development of methodologies for LSCM implementation should take into account these CF, in order to allow a greater adherence of the adopted practices adopted (Karim and Arif-Uz-Zaman 2013; Dora et al. 2016).

5   Final Considerations and Research Direction

Most of the studies published to date on LSCM has focused on outlining practices and their potential benefits, inferring that once companies adopt them the lean implementation would be automatically started. However, such implementation throughout the supply chain is extremely difficult and challenging. The present study aimed to identify the main practices, barriers and contextual factors inherent to the LSCM implementation. To achieve that, a systematic literature review was undertaken, which resulted in a final portfolio with 60 articles. From the analysis of this portfolio, it was verified that most research addresses the three topics (practices, barriers and contextual factors) in an isolated way, not correlating them. Thus, this study contributes to the strengthening of the body of knowledge on LSCM, consolidating 18 practices, 12 barriers and 8 contextual factors inherent to its implementation.

Further, in terms of research agenda, this study found some opportunities for future development due to existing gaps in the literature; they are:

(a) There is still a certain degree of superficiality regarding the understanding of LSCM-specific practices, since many of these are confused with manufacturing practices and do not undergo the adaptations needed to support the complexity of a supply chain. Jasti and Kodali (2015a) comment that the different points of view of the researchers about LSCM results in an accumulation of several incoherent elements, which reveal a deficiency in their standardization used to implement LSCM. This indicates that the LSCM-related theory is still unstable. This research pointed out that few studies have addressed a holistic view of LSCM. Many researchers have focused on analyzing aspects of LSCM practices upstream of the supply chain, while few articles (e.g., Martínez-Jurado and Moyano-Fuentes 2014) have analyzed downstream practices. Therefore, it is proposed the development of studies that identify, classify and validate empirically the main practices of LSCM, in order to direct the construction of a consolidated concept. Moreover, it is suggested studies to analyze the impact that such practices implementation may have on the supply chain performance.

(b) The study consolidated the main barriers inherent to LSCM implementation. However, from the systematic literature review of the literature, it has been emphasized that implementing LSCM practices is not an easy task; there is a gap between theory and practice that raises the question of how to reduce such distance to succeed in LSCM implementation. The difficulty to change behaviors and the lack of commitment of top management were the two most cited barriers in the final portfolio examined. In addition, the absence of an appropriate organizational culture and lack of trust among supply chain agents are vital factors for sustaining LSCM (Jadhav et al. 2014). Given the scarcity of applied research related to the barriers for LSCM implementation (Manzouri et al. 2013), an investigation on their actual effect becomes appropriate. Further, examining the association between these barriers and supply chain contextual factors features an additional opportunity, with both theoretical and practical implications. Such identification allows adopting preventive counter measures to mitigate potential barriers associated with the supply chain context under study, entailing a less problematic LSCM implementation. Thus, investigating the moderating effect of contextual factors on the relationship between barriers and LSCM practices is an opportunity for future studies in the area.

(c) Finally, the incipience of the studies related to the maturity assessment of the supply chains regarding the level of LSCM implementation stands out. The few studies that aimed at assessing maturity of LSCM implementation suggest methods that only approach upstream agents, neglecting the potential downstream relationships. In addition, these methods propose the supply chain evaluation from the perspective of the company under study, leading to possible distortions in the flow analysis. Thus, the determination of methodologies that approach the LSCM implementation as a whole, involving the agents from all tiers, still lacks in the literature. Such gap can even be enlarged if the reverse flow of materials (reverse logistics) is taken into account for implementing LSCM practices and principles. Thereby, further research on LSCM should consider maturity assessment methods that truly evaluate the whole supply chain, and not just a few chunks of it.

References

Adamides, E. D., Karacapilidis, N., Pylarinou, H., & Koumanakos, D. (2008). Supporting collaboration in the development and management of lean supply networks. *Production Planning and Control, 19,* 35–52.

Adebanjo, D., Laosirihongthong, T., & Samaranayake, P. (2016). Prioritizing lean supply chain management initiatives in healthcare service operations: A fuzzy AHP approach. *Production Planning and Control, 27,* 953–966.

Akkermans, H., Bogerd, P., & Van Doremalen, J. (2004). Travail, transparency and trust: A case study of computer-supported collaborative supply chain planning in high-tech electronics. *European Journal of Operational Research, 153,* 445–456.

Al-Aomar, R. (2012). A lean construction framework with six sigma rating. *International Journal of Lean Six Sigma, 3,* 299–314.

Alves Filho, A. G., Cerra, A. L., Maia, J. L., Sacomano Neto, M., & Bonadio, P. V. G. (2004). Pressupostos da gestão da cadeia de suprimentos: Evidências de estudos sobre a indústria automobilística. *Gestão & Produção, 11,* 275–288.

Anand, G., & Kodali, R. (2008). A conceptual framework for lean supply chain and its implementation. *International Journal of Value Chain Management, 2,* 313–357.

Ariffin, A. S., Abas, Z., & Baluch, N. H. (2015). Literature ratified knowledge based view of poultry supply chain integration concept. *Jurnal Teknologi, 77*(27), 35–39.

Arif-Uz-Zaman, K., & Ahsan, A. M. M. N. (2014). Lean supply chain performance measurement. *International Journal of Productivity and Performance Management, 63,* 588–612.

Arkader, R. (2001). The perspective of suppliers on lean supply in a developing country context. *Integrated Manufacturing Systems, 12,* 87–93.

Azevedo, S. G., Carvalho, H., Duarte, S., & Cruz-Machado, V. (2012). Influence of green and lean upstream supply chain management practices on business sustainability. *IEEE Transactions on Engineering Management, 59,* 753–765.

Baker, P. (2004). Aligning distribution center operations to supply chain strategy. *The International Journal of Logistics Management, 15,* 111–123.

Ballou, R. H. (2009). *Gerenciamento da Cadeia de Suprimentos.* Logística Empresarial Bookman Editora.

Behrouzi, F., & Wong, K. Y. (2011). An investigation and identification of lean supply chain performance measures in the automotive SMES. *Scientific Research and Essays, 6,* 5239–5252.

Bevilacqua, M., Ciarapica, F. E., & De Sanctis, I. (2017). Relationships between Italian companies' operational characteristics and business growth in high and low lean performers. *Journal of Manufacturing Technology Management, 28,* 250–274.

Bhamu, J., & Singh Sangwan, K. (2014). Lean manufacturing: Literature review and research issues. *International Journal of Operations & Production Management, 34,* 876–940.

Birgün Barla, S. (2003). A case study of supplier selection for lean supply by using a mathematical model. *Logistics Information Management, 16,* 451–459.

Boonsthonsatit, K., & Jungthawan, S. (2015). Lean supply chain management-based value stream mapping in a case of Thailand automotive industry. In *Proceedings from 4th IEEE International Conference on Advanced Logistics and Transport* (pp. 65–69). IEEE ICALT Institute of Electrical and Electronics Engineers Inc.

Camacho-Miñano, M. D., Moyano-Fuentes, J., & Sacristan-Diaz, M. (2013). What can we learn from the evolution of research on lean management assessment? *International Journal of Production Research, 51,* 1098–1116.

Carvalho, H., Duarte, S., & Machado, V. C. (2011). Lean, agile, resilient and green: Divergencies and synergies. *International Journal of Lean Six Sigma, 2,* 151–179.

Carvalho, H., Govindan, K., Azevedo, S. G., & Cruz-Machado, V. (2017). Modelling green and lean supply chains: An eco-efficiency perspective. *Resources, Conservation and Recycling, 120,* 75–87.

Christopher, M., & Towill, D. (2001). An integrated model for the design of agile supply chains. *International Journal of Physical Distribution & Logistics Management, 31,* 235–246.

Cudney, E., & Elrod, C. (2011). A comparative analysis of integrating lean concepts into supply chain management in manufacturing and service industries. *International Journal of Lean Six Sigma, 2,* 5–22.

Dora, M., Kumar, M., & Gellynck, X. (2016). Determinants and barriers to lean implementation in food-processing SMEs—A multiple case analysis. *Production Planning and Control, 27,* 1–23.

Drohomeretski, E., Gouvea Da Costa, S. E., Pinheiro De Lima, E., & Wachholtz, H. (2012). Lean supply chain management: Practices and performance measuresed. In *Proceedings of 62nd IIE Annual Conference and Expo* (pp. 1869–1880). Orlando, FL: Institute of Industrial Engineers.

Duarte, S., & Cruz Machado, V. (2017). Green and lean implementation: An assessment in the automotive industry. *International Journal of Lean Six Sigma, 8,* 65–88.

Dües, C. M., Tan, K. H., & Lim, M. (2013). Green as the new lean: How to use lean practices as a catalyst to greening your supply chain. *Journal of Cleaner Production, 40,* 93–100.

Eisler, M., Horbal, R., & Koch, T. (2007). Cooperation of lean enterprises—Techniques used for lean supply chain. In J. Olhager & F. Persson (Eds.), *IFIP International Federation for Information Processing* (pp. 363–370). Springer: Boston.

Ellram, L. M., & Cooper, M. C. (2014). Supply chain management: It's all about the journey, not the destination. *Journal of Supply Chain Management, 50,* 8–20.

Erridge, A., & Murray, J. G. (1998). The application of lean supply in local government: The Belfast experiments. *European Journal of Purchasing and Supply Management, 4,* 207–221.

Found, P., Hines, P., Griffiths, G., & Harrison, R. (2008). Creating a sustainable lean business system within a multi-national group company. In *Proceedings of IIE Annual Conference and Expo* (pp. 302–307). Vancouver, BC.

Gueimonde-Canto, A., González-Benito, J., & Manuel García-Vázquez, J. (2011). Competitive effects of co-operation with suppliers and buyers in the sawmill industry. *Journal of Business & Industrial Marketing, 26,* 58–69.

Hadid, W., & Mansouri, S. A. (2014). The lean-performance relationship in services: A theoretical model. *International Journal of Operations & Production Management, 34,* 750–785.

Hartono, Y., Astanti, R. D., & Ai, T. J. (2015). Enabler to successful implementation of lean supply chain in a book publisher. *Procedia Manufacturing, 4,* 192–199.

Hines, P., Holweg, M., & Rich, N. (2004). Learning to evolve: A review of contemporary lean thinking. *International Journal of Operations & Production Management, 24,* 994–1011.

Huang, S. H., Uppal, M., & Shi, J. (2002). A product driven approach to manufacturing supply chain selection. *Supply Chain Management: An International Journal, 7,* 189–199.

Jadhav, J., Mantha, S. S., & Rane, S. B. (2014). Exploring barriers in lean implementation. *International Journal of Lean Six Sigma, 5,* 122–148.

Jajja, M. S. S., Kannan, V. R., Brah, S. A., & Hassan, S. Z. (2016). Supply chain strategy and the role of suppliers: Evidence from the Indian sub-continent. *Benchmarking: An International Journal, 23,* 1658–1676.

Jaklic, J., Trkman, P., Groznik, A., & Stemberger, M. I. (2006). Enhancing lean supply chain maturity with business process management. *Journal of Information and Organizational Sciences, 30,* 205–223.

Jasti, N. V. K., & Kodali, R. (2015a). A critical review of lean supply chain management frameworks: Proposed framework. *Production Planning and Control, 26,* 1051–1068.

Jasti, N. V. K., & Kodali, R. (2015b). Lean production: Literature review and trends. *International Journal of Production Research, 53,* 867–885.

Karim, A., & Arif-Uz-Zaman, K. (2013). A methodology for effective implementation of lean strategies and its performance evaluation in manufacturing organizations. *Business Process Management Journal, 19,* 169–196.

Kumar, S., Luthra, S., Govindan, K., Kumar, N., & Haleem, A. (2016). Barriers in green lean six sigma product development process: An ISM approach. *Production Planning & Control, 27,* 604–620.

La Londe, B. J., Grabner, J. R., & Robeson, J. F. (1971). Integrated distribution systems: A management perspective. *International Journal of Physical Distribution, 1,* 43–49.

Lambert, D. M., & Cooper, M. C. (2000). Issues in supply chain management. *Industrial Marketing Management, 29,* 65–83.

Lambert, D. M., Cooper, M. C., & Pagh, J. D. (1998). Supply chain management: Implementation issues and research opportunities. *The International Journal of Logistics Management, 9,* 1–20.

Lambert, D. M., García-Dastugue, S. J., & Croxton, K. L. (2005). An evaluation of process-oriented supply chain management frameworks. *Journal of Business Logistics, 26,* 25–51.

Lamming, R. (1996). Squaring lean supply with supply chain management. *International Journal of Operations & Production Management, 16,* 183–196.

Lancaster, F. W. (2004). *Indexação e resumos: teoria e prática*. Briquet de Lemos: Tradução de Antônio Agenor Briquet de Lemos. rev. atual Brasília.

Levy, D. L. (1997). Lean production in an international supply chain. *Sloan Management Review, 38,* 94.

Li, S., Ragu-Nathan, B., Ragu-Nathan, T., & Rao, S. S. (2006). The impact of supply chain management practices on competitive advantage and organizational performance. *Omega, 34,* 107–124.

Machado, V. C., & Pereira, A. (2008). Modelling lean performance. In *Proceedings of 4th IEEE International Conference on Management of Innovation and Technology* (pp. 1308–1312). Bangkok: ICMIT.

Manzouri, M. (2012). How lean supply chain implementation affect halal food companies. *Advances in Natural and Applied Sciences, 6,* 1485–1489.

Manzouri, M., Ab-Rahman, M. N., Zain, C. R. C. M., & Jamsari, E. A. (2014). Increasing production and eliminating waste through lean tools and techniques for Halal food companies. *Sustainability (Switzerland), 6,* 9179–9204.

Manzouri, M., Nizam Ab Rahman, M., Saibani, N., & Rosmawati Che Mohd Zain, C. (2013). Lean supply chain practices in the Halal food. *International Journal of Lean Six Sigma, 4,* 389–408.

Manzouri, M., & Rahman, M. N. A. (2013). Adaptation of theories of supply chain management to the lean supply chain management. *International Journal of Logistics Systems and Management, 14,* 38–54.

Marodin, G. A., Frank, A. G., Tortorella, G. L., & Saurin, T. A. (2016). Contextual factors and lean production implementation in the Brazilian automotive supply chain. *Supply Chain Management: An International Journal, 21,* 417–432.

Martínez-Jurado, P. J., & Moyano-Fuentes, J. (2014). Lean management, supply chain management and sustainability: A literature review. *Journal of Cleaner Production, 85,* 134–150.

Mcivor, R. (2001). Lean supply: The design and cost reduction dimensions. *European Journal of Purchasing and Supply Management, 7,* 227–242.

Morgan, C. (2007). Supply network performance measurement: Future challenges? *The International Journal of Logistics Management, 18,* 255–273.

Olesen, P., Powell, D., Hvolby, H.-H., & Fraser, K. (2015). Using lean principles to drive operational improvements in intermodal container facilities: A conceptual framework. *Journal of Facilities Management, 13,* 266–281.

Paré, G., Trudel, M.-C., Jaana, M., & Kitsiou, S. (2015). Synthesizing information systems knowledge: A typology of literature reviews. *Information & Management, 52,* 183–199.

Parveen, C. M., Kumar, A. R. P., & Narasimha Rao, T. V. V. L. (2011). Integration of lean and green supply chain—Impact on manufacturing firms in improving environmental efficiencies. In *Proceedings of International Conference on Green Technology and Environmental Conservation* (pp. 143–147). Chennai: GTEC.

Parveen, M., & Rao, T. V. V. L. N. (2009). An integrated approach to design and analysis of lean manufacturing system: A perspective of lean supply chain. *International Journal of Services and Operations Management, 5,* 175–208.

Perez, C., De Castro, R., Simons, D., & Gimenez, G. (2010). Development of lean supply chains: A case study of the Catalan pork sector. *Supply Chain Management: An International Journal, 15,* 55–68.

Professionals, C. O. S. C. M. (2013). *CSCMP supply chain management definitions and glossary* [online]. Available at: http://cscmp.org/CSCMP/Educate/SCM_Definitions_and_Glossary_of_Terms/CSCMP/Educate/SCM_Definitions_and_Glossary_of_Terms.aspx?hkey=60879588-f65f-4ab5-8c4b-6878815ef921. Accessed July 19, 2017.

Qi, Y. N., & Chu, Z. F. (2009). The impact of supply chain strategies on supply chain integration. In *Proceedings of 16th International Conference on Management Science and Engineering* (pp. 534–540). Moscow: ICMSE.

Qrunfleh, S., & Tarafdar, M. (2013). Lean and agile supply chain strategies and supply chain responsiveness: The role of strategic supplier partnership and postponement. *Supply Chain Management: An International Journal, 18*, 571–582.

Rahman, S., Laosirihongthong, T., & Sohal, A. S. (2010). Impact of lean strategy on operational performance: A study of Thai manufacturing companies. *Journal of Manufacturing Technology Management, 21*, 839–852.

Shah, R., & Ward, P. T. (2003). Lean manufacturing: Context, practice bundles, and performance. *Journal of Operations Management, 21*, 129–149.

Sharma, V., Dixit, A. R., & Qadri, M. A. (2015). Impact of lean practices on performance measures in context to Indian machine tool industry. *Journal of Manufacturing Technology Management, 26*, 1218–1242.

Soni, G., & Kodali, R. (2016). Interpretive structural modeling and path analysis for proposed framework of lean supply chain in Indian manufacturing industry. *Journal of Industrial and Production Engineering, 33*, 501–515.

Sridharan, U. V., Royce Caines, W., & Patterson, C. C. (2005). Implementation of supply chain management and its impact on the value of firms. *Supply Chain Management: An International Journal, 10*, 313–318.

Stavrulaki, E., & Davis, M. (2010). Aligning products with supply chain processes and strategy. *International Journal of Logistics Management, 21*, 127–151.

Taylor, D. H. (2006). Strategic considerations in the development of lean agri-food supply chains: A case study of the UK pork sector. *Supply Chain Management: An International Journal, 11*, 271–280.

Tortorella, G. L., Marodin, G. A., Miorando, R., & Seidel, A. (2015). The impact of contextual variables on learning organization in firms that are implementing lean: A study in Southern Brazil. *The International Journal of Advanced Manufacturing Technology, 78*, 1879–1892.

Ugochukwu, P., Engström, J., & Langstrand, J. (2012). Lean in the supply chain: A literature review. *Management and Production Engineering Review, 3*, 87–96.

Vilaça, M. L. C. (2010). Pesquisa e ensino: Considerações e reflexões. *Revista do Curso de Letras da UNIABEU, 1*, 59–74.

Vitasek, K. L., Manrodt, K. B., & Abbott, J. (2005, October). What makes a lean supply chain? *Supply Chain Management Review, 9*(7), 39–45.

Vlachos, I. (2015). Applying lean thinking in the food supply chains: A case study. *Production Planning & Control, 26*, 1351–1367.

Wee, H. M., & Wu, S. (2009). Lean supply chain and its effect on product cost and quality: A case study on Ford Motor Company. *Supply Chain Management: An International Journal, 14*, 335–341.

Wiengarten, F., Fynes, B., & Onofrei, G. (2013). Exploring synergetic effects between investments in environmental and quality/lean practices in supply chains. *Supply Chain Management: An International Journal, 18*, 148–160.

Womack, J. P., & Jones, D. T. (1996). *Lean thinking: Banish waste and create wealth in your organisation* (p. 397). New York, NY: Simon and Shuster.

Yew Wong, C., Stentoft Arlbjørn, J., & Johansen, J. (2005). Supply chain management practices in toy supply chains. *Supply Chain Management: An International Journal, 10*, 367–378.

A Literature Review on Lean Manufacturing in Small Manufacturing Companies

Laís Ghizoni Pereira and Guilherme Luz Tortorella

Abstract This chapter aims to identify, through a systematic literature review, the main Lean Manufacturing (LM) practices, critical success factors (CSF) and barriers within small manufacturing companies' context. This paper presents a systematic literature review based on the proposed approach denoted as ProKnow-C to identify the correlated bibliographic portfolio (BP). Our findings indicate that the consolidation of specific CSF related to the context of small manufacturing companies reinforces the body of knowledge, reinforcing the establishment of a broader perspective of LM implementation in these companies. Further, the capability of disseminating the continuous improvement mindset across all employees is a significant challenge for these companies, since their leaders are poorly trained in accordance with the underlying LM principles. The recent growth of small companies and their relevance to socioeconomic development has raised the importance of improving their management processes. Particularly for LM implementation, few studies have specifically approached this context whose challenges may be differentiated, highlighting the need for a better comprehension of proper practices, barriers, and CSF.

Keywords Small companies · Lean manufacturing · Literature review

1 Introduction

To remain competitive, companies need to improve several aspects such as costs, quality, delivery service, and flexibility, which motivate changes in current manufacturing systems (Shah and Ward 2003). Among the existing improvement

L. G. Pereira · G. L. Tortorella (✉)
Department of Systems and Industrial Engineering, Universidade Federal de Santa Catarina, Florianópolis, Brazil
e-mail: gtortorella@bol.com.br

L. G. Pereira
e-mail: laisghizoni@hotmail.com

© Springer International Publishing AG, part of Springer Nature 2018
J. P. Davim (ed.), *Progress in Lean Manufacturing*,
Management and Industrial Engineering,
https://doi.org/10.1007/978-3-319-73648-8_3

approaches, Lean Manufacturing (LM) is widely deemed as an organizational philosophy that aims to systematically eliminate waste and add value to customers (Womack and Jones 1992; Ohno 1998; Matt 2008). The implementation of LM has been gaining importance due to the benefits observed in several areas and sectors (Chen and Meng 2010). Thus, LM emerges as an approach that provides ways to improve quality, meet customer expectation, reduce waste in all forms, enhance employees' satisfaction and shorten the delivery times (Braglia et al. 2006; Bakas et al. 2011).

Although widely disseminated, LM practices and principles are not easily implemented (Treville and Antonakis 2006). Specifically for small-sized companies, due to lower degrees of process standardization and short-term management, these changes may require an adapted approach for allowing the continuous improvement implementation (Antony et al. 2005; Achanga et al. 2006; Bakas et al. 2011). According to Shah and Ward (2003), companies' context must be taken into account when implementing LM. Among the contextual variables, the size of the company (usually denoted by the number of employees) stands out as a key variable for influencing the level of adoption of LM practices and principles. Additionally, previous studies (e.g., Achanga et al. 2006; Kumar et al. 2009; Nordin et al. 2010; Dora et al. 2013; Marodin and Saurin 2015b; Saurin et al. 2010) have also addressed the effect of the companies' size on the LM implementation. Generally, these studies indicate that larger companies are more likely to fully implement LM than smaller companies.

Further, small-sized companies have a significant relevance in the global socioeconomic scenario, representing more than half of the existing companies (Antony et al. 2005). Besides, such companies play a vital role in the economic growth of developing countries, accounting for more than 90% of commercial establishments and about 50% of exports. The availability of jobs is also affected, since they provide the largest amount of opportunities, reaching 70% of the workforce in Europe (Yang and Yu 2010) and 50% in North America (CEC 2005). In Brazil, where small companies represent 25% of GDP—Gross Domestic Product, their importance is also noticed. In addition, small companies (employing up to 100 employees) comprise 99% of formal establishments and employ 52% of the workforce in the country (SEBRAE 2014).

Therefore, LM implementation can help these companies to achieve more striking results, by increasing their competitiveness (Antony et al. 2005; Achanga et al. 2006). Due to the inherent difficulties of LM implementation in small companies, it is important to provide a deeper understanding of the aspects that effectively influence its adoption under this context. Based on these arguments, the following research question is raised: *what are the critical success factors (CSF), barriers, and practices for Lean Manufacturing implementation in small manufacturing companies?*

Thus, this article aims to identify based on a systematic literature review the main critical success factors, barriers, and practices for LM implementation in small manufacturing companies. The proposed method is grounded in the application of the Knowledge Development Process—Constructivist (ProKnow-C), which presents a methodology for the construction of scientific knowledge based on a bibliographic portfolio (BP) relevant to the theme (Dutra et al. 2015; Ensslin et al. 2015; Lacerda et al. 2012). Besides the contribution to the body of knowledge, this study provides future research directions to fulfill existing gaps in the literature.

2 Research Method

According to Paré et al. (2015), the literature review is the main method for reinforcing a research problem, justifying the proposed approach according to the existing knowledge and research gaps (Cardoso et al. 2015). This research has an exploratory characteristic aiming to generate knowledge about CSF and barriers for LM implementation in small companies, highlight opportunities and research directions. The study relies on the collection of primary and secondary data. Primary data are obtained directly from the delimitations imposed by the researchers in the databases' search; whereas secondary data are derived from the resultant BP. Thus, the structured literature review process namely as ProKnow-C was (Ensslin et al. 2014; Dutra et al. 2015), as detailed below.

As stated by Cardoso et al. (2015) and Ensslin et al. (2015), several studies have applied ProKnow-C to: (i) identify a BP on a given subject; (ii) identify studies characteristics in the area of knowledge; (iii) perform a BP critical reading; and finally, (iv) identify a research question that characterizes an opportunity for future work. The composition of a BP is characterized by a restricted group of publications with scientific recognition, which is selected by the researcher according to predetermined criteria (Cardoso et al. 2015).

The first stage of the BP selection consisted in the definition of the research axes. Although this research focuses on small companies, the selection of BP was made by searching for articles within all contexts, allowing comparisons within the scientific scope. To achieve that, two axes were defined: (i) critical factors/barriers/best practices and (ii) lean manufacturing. Therefore, the keywords presented in Table 1 were related to retrieve the publications in the titles, abstracts, and/or keywords. Scientific articles were identified by keywords in the following databases: Web of Knowledge (ISI), Science Direct, Engineering Village, Scopus, EBSCO, and ProQuest. The results of the initial search presented 2272 articles, as shown in Table 1.

In this step, the "keyword adherence test" was applied to validate the keywords used in the initial search. Five articles were randomly selected from the 2272 previously identified, and their keywords compared to those used in the research axes, as recommended by Ensslin et al. (2013). Based on this comparison, it was observed that the used keywords were present in the set of keywords of the articles, making it unnecessary to incorporate additional ones. As for the filtering process, we used the EndNote® X7 and Google Scholar. This step began with the elimination of repeated articles and then a sequence of analyzes aiming to verify the alignment with the investigated subject, such as appropriateness of the titles,

Table 1 Database and keywords

Databases	Keywords			Total publications
Web of knowledge	*"Critical factors" or "Barrier*" or "Performance measure*" or "Best practices"*	And	*"Lean manufacturing" or "Lean production" or "Lean system"*	145
Science direct				63
Engineering village				157
Scopus				813
EBSCO				104
ProQuest				990
Total				2272

scientific relevance, abstracts, and availability of complete articles. Finally, the full reading of the articles allowed the verification of complete alignment with the research subject.

From the 2272 articles identified from the databases, it was noticed that 691 articles were repeated or were not characterized as Journal Article or Conference Proceedings; entailing at their disposal. The remaining 1581 non-repeated articles were analyzed regarding the alignment of their titles; of which only 149 were in line with the research subject. Subsequently, these 149 articles were analyzed according to their scientific relevance, through their citations frequency obtained in Google Scholar. A threshold value for the accumulated citation of 95% was established and the articles that added up to this value were selected, resulting in 48 articles (grouped as repository K). Thus, the abstracts of these articles were assessed as for their alignment with the research subject, only 27 being selected. The authors of these 27 articles comprised the "Base of Authors" (BA) composed of 65 researchers. The remaining 101 articles, which represented 5% of the total citations, were allocated to another group named repository P.

As for the articles in "P", an analysis was made regarding the period of publication. Articles published more than 2 years ago by authors included in the BA were maintained and added to articles published less than 2 years ago, totaling 45 articles. These articles' abstracts were analyzed for their alignment with the research subject, resulting in 24. Hence, the 24 articles selected from the P repository were added to the 27 articles in the K repository, totaling 51 articles. From this, 5 articles did not present their full texts and after reading them, only 41 articles were fully aligned with the research subject in their titles, abstracts, and texts, hence proceeding to the next step of ProKnow-C.

The representativeness test of the 41 articles was performed based on the total references listed in these articles, which totaled 1973 references. Similarly, duplicate references and those whose year of publication is prior to 2000 were eliminated. In addition, only references from scientific journals or conference proceedings were considered, which led to the selection of 483 articles. After analyzing the alignment of their titles and abstracts, 39 articles were selected from which 29 were already included in the BP, entailing the read of 10 articles. Out of these, only 3 articles were aligned and added to the final BP. Thus, in total, 44 papers were allocated to the final BP (see Appendix), from which the gaps and research opportunities were identified.

After defining the BP, we performed a content analysis aiming at consolidating the existing knowledge with respect to the research subject (Ensslin et al. 2014). Therefore, some characteristics of the BP were selected in order to identify their occurrences and provide theoretical arguments (Ensslin et al. 2013; Cardoso et al. 2015). This content analysis was divided into two types of variables: basic and advanced. The basic variables comprise the researchers with publication trajectory in the area of knowledge, and the articles with the highest scientific recognition, as suggested by Dutra et al. (2015). Additionally, it is also possible to incorporate the variable "temporal evolution" which identifies the periods of significant change in relation to the research subject. The advanced variables are composed of: (i) characteristics of small companies that implement LM; (ii) main LM practices implemented in this context; (iii) CSF; and (iv) barriers to LM implementation in small companies. Based on the 44 BP articles, only 24 out of the 79 identified authors presented at least 2 articles published in BP (see Table 2). It is worth noting that *Jiju Antony, Maneesh Kumar,* and *Giuliano Marodin* have authored more than 5 publications. As for the journals and conferences included in the BP, the *International Journal of Production Research* stands out as the one with the highest amount of publications. Regarding the period of publication, during 2009 and 2010 there was a significant increase in the number of articles published on the subject. Publications from this period are mostly from Europe, which might be justified by the growth of the representativeness of small and medium—in the United Kingdom and other European countries (Antony et al. 2005; Dora et al. 2013). It is worth to notice that 8 articles (approximately 20% of BP) were published during the last 2 years, indicating the recent importance of the subject.

With regards to the advanced variables, the articles were categorized according to: (i) small companies' context, (ii) CSF for LM implementation, (iii) barriers, and (iv) description or citation of LM practices. For the small companies' context, 80% of articles emphasize the need to evaluate companies individually, given the influence that company size can present on a successful LM implementation (e.g., Shah and Ward 2003; Kumar et al. 2009). For the CSF, more than half of the BP articles mentions them explicitly, emphasizing the importance of addressing organizational issues prior to or in parallel with LM implementation (e.g., Yew Wong 2005; Achanga et al. 2006; Bakas et al. 2011). Regarding the barriers to implement LM, 43% of the articles carry out their identification in order to reinforce or complement the knowledge about the existing challenges (e.g., Saurin et al. 2010; Bhasin 2012a; Abolhassani et al. 2016). Finally, 59% of the articles present the main LM practices. Overall, only 2 out of 44 articles approached all advanced variables concurrently, and only 18 presented at least 3 out of the 4 variables. Such finding indicates the scarcity of studies that address more broadly the LM implementation within small companies' context.

Table 2 Number of publications by author, journal/conference and year

Author	Total publications	Journal/Event	Total publications
Antony, J.	6	Int. J. of Production Research	5
Kumar, M.	6	J. of Manufacturing Technology Manag.	4
Marodin, G.	5	Int. J. of Innovation, Management and Tech.	2
Deros, Md. B.	4	Int. J. of Operations & Production Manag.	2
Kodali, R.	3	Int. J. of Productivity and Performance Manag.	2
Nordin, N.	3	Int. J. of Quality & Reliability Manag.	2
Saurin, T.	3	J. of Advanced Manufacturing Systems	2
Wong, K.	3	J. of Operations Manag.	2
Anand, G.	2	J. of the Operational Research Society	2
Bhasin, S.	2	Manag. Decision	2
Dombrowski, U.	2	Other 19 journals/events	19
Doolen, T.	2	**Year**	**Total publications**
Dora, M.	2	2001	1
Gellynck, X.	2	2003	1
Molar, A.	2	2005	3
Rahman, M.	2	2006	3
Rose, A.	2	2007	2
Roy, R.	2	2008	1
Shah, R.	2	2009	7
Tortorella, G.	2	2010	6
Van Goubergen, D.	2	2011	2
Van Landeghem, H.	2	2012	5
Wahab, D.	2	2013	2
Ward, P.	2	2014	3
Other 55 authors (each)	1	2015	4
		2016	4

3   Characterization of Small Companies in LM Implementation

Companies' size is quite discussed in the literature, with several ways of approaching it, both quantitative and qualitatively. Quantitative classifications generally consider either the company's number of employees or the annual turnover, since they are measurable and the size delimitation is of ease identification. On the other hand, qualitative classifications are established based on aspects such as managers' behavioral criteria, style of the management and interaction with the customers and suppliers (Leone 1991). Regardless of the classification form, companies' size can be divided into large, medium, small, and micro, and may vary according to the general conditions of each country, region, or organization. In developed countries, for instance, some small companies would be considered medium- or even large-sized in nonindustrialized countries. Further, in countries with larger territorial extension and/or with significant socioeconomic unbalances, such as Brazil, the same situation can also occur. In general, there is no widely accepted definition, as some countries do not distinguish between small and medium-sized companies (Bakas et al. 2011).

Antony et al. (2005) claim that smaller companies are vital for modern economies. In some European countries, for instance, they account for more than 60% of all companies and employ most of the labor force, contributing for value creation (annual turnover estimated at US$1.3 trillion) (Dora et al. 2013). In North America, the economic importance of small companies is also perceived, representing more than 98% of Canadian, Mexican, and United States companies distributed in several segments, and producing about 40% of their Gross Domestic Product (GDP) (CEC 2005).

The participation of small companies in Brazil has been growing over the years and has had positive results. In 2001, their participation in the Brazilian GDP was 23.2% and, in 2011, reached 27% (SEBRAE 2014). Moreover, in 2015, small companies' export increased 7.5% compared to 2014, with 142 new small companies launched in the international market even under the context of external crisis (APEXBRASIL 2015). However, the mortality of small manufacturing companies before the first 2 years is close to 20.1% (SEBRAE 2013).

In terms of resources, small companies face significant limitations. Usually, they are limited not only in human resources (due to low investments in training and qualification), but also in financial capability to improve infrastructure. Therefore, such limitations may undermine a feasibility of any longer term planning, compromising the companies' development and growth (Antony et al. 2005; Kumar et al. 2009; Dora et al. 2013).

Further, the organizational structure of small companies is often flat and less hierarchical than in larger companies. It is also more flexible, making it easier to implement new methods and promote change, avoid bureaucracy, and create a positive work environment with higher satisfaction. However, the informal atmosphere impairs processes standardization, jeopardizing the implementation and sustainability of several LM practices (Timans et al. 2012; Dora et al. 2013; Marodin and Saurin 2015a, b).

4 LM Practices in Small Companies

One way to define the steps for implementing LM is to identify the appropriate practices for the aimed context. LM practices can be found in strategic, tactical, and operational levels within the companies, regardless their size. This identification allows to establish the priorities with respect to the improvement efforts (Anand and Kodali 2010; Abolhassani et al. 2016; Belhadi and Touriki 2016). In this sense, some practices may require greater capital expenditure, which eventually complicates their implementation in small companies with lower financial resources (Dora et al. 2014). Thus, managers tend to allocate specific teams to facilitate the implementation of such practices, developing low-cost and high-impact solutions (Kumar et al. 2014; Godinho Filho et al. 2016).

According to the investigated literature, the most evidenced LM practice in small companies is p_1 (pull production), as shown in Table 3. These companies usually have area restrictions, which tends to limit the availability for storage. Therefore, to produce according to customers' demand becomes essential due to mitigate products inventory (Matt and Rauch 2013; Abolhassani et al. 2016; Belhadi and Touriki 2016). Moreover, they tend to present larger restrictions on their cash flow, which is directly influenced by the amount of inventory (Tortorella et al. 2015b). Since the implementation of pull production can lead to reduced inventories, companies' cash flow might benefit from it, allowing greater flexibility of the business (Dora et al. 2013).

The second most cited practice, p_2 (Total Productive Maintenance), was conceived with an initial focus on machine maintenance processes, and broadened its scope to other supporting processes (Zhou 2016). This practice aims to eliminate losses generated in the material flow through the integration of the maintenance and production, as well as to prevent failures during the process (Dora et al. 2014). The expectation from the adoption of this practice is to increase process stability, which is a fundamental issue for LM implementation. Such stability can also be achieved through the adoption of practices p_3 (Kaizen) and p_4 (5S), which encourage continuous improvement activities and reinforce the implementation of simple ideas led by the natural work teams (Doolen and Hacker 2005; Dombrowski et al. 2010).

On the other hand, practice p_{19}—PDCA (Plan, Do, Check, Act) presents lowest citation frequency in the researched literature. This practice is characterized by assisting in process management and problem-solving, allowing adequate analysis, planning, and reflection on proposed improvement activities (Rose et al. 2013; Dora et al. 2014). Wong et al. (2009) suggests that this practice is more usually evidenced in companies that have implemented LM for more than 10 years, due to the complexity of verifying the assertiveness of PDCA cycles in the shorter term. Further, many small companies, with limited resources, end up focusing on the implementation of operational practices instead of managerial ones (Dombrowski and Mielke 2014; Tortorella et al. 2015a). According to Wong et al. (2009), this practice is more frequently associated with longer term improvement activities.

Table 3 Main LM practices identified in BP

LM practices	Authors																											Total occurrences in BP
	1	3	4	6	7	10	11	12	13	15	16	17	18	19	20	24	25	30	31	32	33	34	37	39	40	41	44	
p_1—Pull system production	X			X	X	X		X	X	X				X	X		X	X		X		X			X	X	X	16
p_2—Total productive maintenance	X			X	X		X	X	X	X				X	X			X	X	X	X	X			X		X	16
p_3—Kaizen/continuous improvement		X		X			X						X	X	X	X	X	X	X	X		X			X	X	X	15
p_4—5S				X	X	X	X	X						X	X		X	X			X				X	X	X	13
p_5—Just in time					X			X		X				X	X	X	X	X	X		X				X		X	12
p_6—Cycle time reduction								X		X		X			X	X	X	X	X		X	X						10
p_7—Total quality management			X		X			X			X	X	X	X				X		X		X						10
p_8—Cellular manufacturing				X	X			X						X	X		X	X		X		X						9
p_9—Quick changeover	X				X	X		X								X	X	X								X	X	9
p_{10}—People empowerment		X											X		X	X		X				X		X		X		8
p_{11}—Value stream mapping				X	X		X								X	X				X						X		7
p_{12}—Standardized work								X			X		X			X		X		X					X			7
p_{13}—Six sigma				X								X	X	X								X					X	6
p_{14}—Cross functional team													X		X	X				X		X	X					6

(continued)

Table 3 (continued)

LM practices	Authors																											Total occurrences in BP
	1	3	4	6	7	10	11	12	13	15	16	17	18	19	20	24	25	30	31	32	33	34	37	39	40	41	44	
p_{15}—Statistical process control			X	X	X					X	X																	5
p_{16}—Visual management		X															X	X		X							X	5
p_{17}—Continuous flow	X									X											X	X					X	5
p_{18}—Poka-yoke (error proofing)	X		X														X		X								X	5
p_{19}—PDCA						X													X						X		X	4

Surprisingly, researchers such as Rose et al. (2013) and Worley and Doolen (2006) approach p_{19} independently of p_3, considering it to be inherent in kaizen activities.

5 Critical Success Factors

The identification of CSF is important to assist the management to succeed in LM implementation. Hence, if these CSF are not properly established within the organization, the implementation is more likely to fail (Yew Wong 2005). Previous studies (e.g., Antony et al. 2005, 2006; Kumar et al. 2009) reinforce the importance of understanding the relationship between CSF and lean implementation, highlighting their role in the context of small companies.

From the consolidation displayed in Table 4, 15 CSF were identified based on the BP; the CSF most cited is f_1—Leadership. According to Abolhassani et al. (2016), small companies' managers generally attribute lean implementation failures to employees' resistance to change. However, previous studies (e.g., Achanga et al. 2006; Kumar et al. 2009) indicate that the responsibility for conducting employees towards an LM transformation relies on the leaders, who should lapidate the behavioral change inherent to a lean implementation. Leadership training could provide a better understanding of its importance in the LM implementation, entailing behavioral changes that support leading through own example (Yew Wong 2005; Netland 2016). Specifically, in the context of small companies, these leaderships tend to accumulate operational activities, overburdening them with routine activities, and reducing their readiness to return to medium and long-term improvement planning (Antony et al. 2005; Bakas et al. 2011). Thus, it is reasonable to expect that the adequate development of small companies' leaders will be a prominent CSF for the LM implementation.

The second most cited CSF, f_2—Culture, is usually claimed to be correlated with f_1. When leaders understand their role in demonstrating and orienting behaviors coherent LM principles, concurrently they begin to influence the organizational culture (Bhasin 2012b; Godinho Filho et al. 2016). Specifically, in small companies' context, the owners are usually part of the senior management, which reinforces their influence on organizational culture (Achanga et al. 2006). However, this CSF can also be influenced by other complementary contextual aspects, such as typical habits and costumes of the region in which the company is located (Netland 2016), socioeconomic factors (e.g., emerging or developed economies countries) (Tortorella et al. 2015b), and the industrial sector to which the company belongs (Yew Wong 2005). The weighting of these contextual aspects can either culminate in behaviors that converge to lean principles, or hinder the construction of an appropriate organizational culture.

On the other hand, the CSF with the lowest citation frequency is f_{13} (organizational infrastructure). The literature on it presents ambiguous definitions that allow different interpretations of its importance for LM implementation in small companies. Timans et al. (2012), for instance, associate this CSF with the way

Table 4 CSF identified in BP

CSF	Authors																									Total occurrences in BP
	2	3	4	5	6	7	8	9	10	11	14	15	16	17	18	19	21	25	26	27	32	36	38	40	41	
f_1—Leadership	X	X	X	X	X	X			X	X	X	X	X	X	X	X			X	X					X	17
f_2—Culture	X	X		X	X	X	X		X		X	X		X	X	X				X	X	X		X		16
f_3—Skills	X	X	X			X		X	X		X	X		X	X	X	X	X								13
f_4—Strategy alignment all levels			X	X	X		X		X	X		X		X	X	X	X									11
f_5—Education and training			X	X	X									X	X	X	X		X	X		X			X	11
f_6—Senior management commitment	X	X	X		X							X		X			X					X		X	X	10
f_7—Communication				X						X		X		X	X	X				X	X	X			X	10
f_8—Stakeholders involvement				X	X			X	X			X		X	X	X						X			X	10
f_9—Resource allocation	X			X	X	X					X	X						X	X						X	9
f_{10}—Continuous learning							X	X	X				X						X	X	X	X	X			9
f_{11}—Commitment/Motivation of workers		X		X	X										X	X	X		X						X	8
f_{12}—Performance evaluation (long-term focus)				X		X			X					X					X			X			X	7
f_{13}—Organizational infrastructure															X	X						X				3

departments are organized and the definition of their respective assignments, in order to structure the employees' role and the hierarchy within the company. Further, Kumar et al. (2009, 2014) emphasize the importance of this CSF in providing a physical structure that supports LM implementation. In general, smaller companies tend to present very limited availability of resources, as aforementioned. Such limitation may differentiate or even catalyze the LM success. Thus, the divergence found in the studies included in the BP regarding the definition of this CSF can influence the understanding of its importance, justifying its low frequency of citation. Overall, the lack of emphasis on these CSF entails the additional challenges for the company, which can impair the LM implementation. However, it is worth noticing that some barriers, complementary to the absence of the aforementioned CSF, must be emphasized.

6   Barriers

While there are several aspects of small companies that provide a suitable environment for a successful LM implementation, there are also obstacles and barriers that must be addressed (Achanga et al. 2006; Yang and Yu 2010). Table 5 lists the barriers evidenced in the BP within the small manufacturing companies' context. Their identification complements the lack of emphasis given by the companies to the aforementioned CSF.

The most frequently cited barrier for these companies was b_1 (lack of benefits understanding), which may occur due to both technical and/or sociocultural misunderstanding, as well as the potential results of the implementation (Kumar et al. 2014; Marodin and Saurin 2015a, b). Although LM benefits are widely acknowledged and well evidenced in several industrial sectors, the obtained gains at the beginning of implementation may conflict with the long-term ones, which are achieved through the construction of a continuous improvement culture (Bakas et al. 2011; Jadhav et al. 2014). Such an apparent paradox can lead to confusion and misinterpretation of the real benefits inherent to LM (Saurin et al. 2010). Thus, it is reasonable to expect that this barrier appears as the most cited one.

Small companies tend to have less bureaucracy and a closer relationship between leadership and operation, which in some cases may be positive for addressing the continuous improvement activities. However, this informal flow of information can lead to hasty interventions without proper planning, entailing the failure of the improvement initiative. Hence, employees may perceive it and misunderstand LM implementation (Zhou 2016), hindering their acceptance and willingness to contribute to new improvement initiatives. Therefore, although the barrier b_4 (Failure of previous improvement projects) presents a low citation in the BP, it has great practical implication for LM implementation, since it may generate disinterest and lack of motivation of the employees (Saurin et al. 2010).

Table 5 Barriers identified in BP

Barriers	Authors																				Total occurrences in BP
	1	3	5	7	9	13	15	17	18	19	22	23	27	28	29	32	35	39	43	44	
b_1—Lack of understanding of benefits	X		X	X	X		X			X	X	X	X	X	X	X		X		X	14
b_2—Employees' resistance to change	X	X	X		X	X			X	X				X	X	X			X	X	12
b_3—Difficulty of adapting improvement projects			X					X								X			X		4
b_4—Failure of previous improvement projects																X				X	2

7 Final Considerations

Due to an increasing relevance of small companies, whose influence has been gaining economic expression both in the developed and emerging countries' context, improvements on the efficiency and productivity of their processes become essential to meet market demands and assure competitiveness. Thus, the adoption of LM principles and practices can support the identification and elimination of wastes, culminating in the reduction of operating costs. However, these companies must be able to identify which LM practices are appropriate to their contexts. Additionally, the emphasis on CSF for LM implementation should be addressed in order to ensure the construction of a favorable culture for sustaining the improvement approach.

Hence, this study aimed to identify the main practices, CSF and barriers for LM implementation in small manufacturing companies through a systematic literature review. To achieve that, we applied an existing literature review method denoted as ProKnow-C. The contribution of this literature review is twofold. First, the consolidation of specific CSF related to the context of small manufacturing companies reinforces the body of knowledge, highlighting the need for establishing a broader perspective of LM implementation in these companies. In addition, among the listed practices, there are different levels of difficulty whose identification may be important given the low availability of resources allocation. Second, the capability of disseminating the continuous improvement mindset across all employees, inserting it in the daily routine processes, is a significant challenge for lean implementation. The systematic literature review indicated that LM implementation in many companies can fail due to the absence of trained and committed leaders in accordance with the underlying principles. Therefore, the consolidation of previous research provides evidence for assuming that leadership and organizational culture are the primary CSF for lean implementation, deserving a differentiated emphasis from managers and practitioners.

Regarding future research opportunities, the study reveals that there are still gaps related to lean implementation in small companies. As a result, it is important to highlight two directions for future research:

(i) Application of a holistic approach to LM implementation in small companies: most of the existing evidence focused on small companies is based on case studies limited to the implementation of lean practices separately. In this sense, BP articles present a superficial relationship between practices, CSF and the barriers to lean implementation. Thus, studies that incorporate practices, CSF and barriers in small companies undergoing an LM implementation are scarce, and may lead to a deeper understanding of the existing interactions in this context. Hence, further research is suggested in the exclusive context of small companies and how this influences lean implementation; and

(ii) Establishing a framework for lean implementation in small companies: liter-
      ature evidence that structures LM implementation with a focus on small
      manufacturing companies is still rare. Moreover, the few studies found suggest
      a generic and superficial set of actions, causing the need for interpretation by
      top management and hindering its proper application. Thus, in addition to its
      theoretical contribution, research that proposes a framework for systematizing
      the implementation process of LM in small companies brings significant
      practical implications, particularly for these companies' leaders who are
      generally less prepared in both technical and sociocultural aspects.

References

Abolhassani, A., Layfield, K., & Gopalakrishnan, B. (2016). Lean and US manufacturing industry:
    Popularity of practices and implementation barriers. *International Journal of Productivity and
    Performance Management, 65*(7), 875–897.
Achanga, P., Shehab, E., Roy, R., & Nelder, G. (2006). Critical success factors for lean
    implementation within SMEs. *Journal of Manufacturing Technology Management, 17*(4),
    460–471.
Anand, G., & Kodali, R. (2010). Analysis of lean manufacturing frameworks. *Journal of Advanced
    Manufacturing Systems, 9*(1), 1–30.
Antony, J., Kumar, M., & Madu, C. (2005). Six sigma in small-and medium-sized UK
    manufacturing enterprises: Some empirical observations. *International Journal of Quality &
    Reliability Management, 22*(8), 860–874.
APEXBRASIL. (2015). Relatório de gestão exercício 2015. Agência Brasileira De Promoção De
    Exportações E Investimentos Apex-Brasil.
Bakas, O., Govaert, T., & Van Landeghem, H. (2011). Challenges and success factors for
    implementation of lean manufacturing in European SMES. In *13th International conference on
    the Modern Information Technology in the Innovation Processes of the Industrial Enterprise
    (MITIP 2011)* (Vol. 1). Tapir Academic Press.
Belhadi, A., & Touriki, F. (2016). A framework for effective implementation of lean production in
    Small and medium-sized enterprises. *Journal of Industrial Engineering and Management, 9*(3),
    786–810.
Bhasin, S. (2012a). An appropriate change strategy for lean success. *Management Decision, 50*(3),
    439–458.
Bhasin, S. (2012b). Prominent obstacles to lean. *International Journal of Productivity and
    Performance Management, 61*(4), 403–425.
Braglia, M., Carmignani, G., & Zammori, F. (2006). A new value stream mapping approach for
    complex production systems. *International Journal of Production Research, 44*(18–19), 3929–
    3952.
Cardoso, T., Ensslin, S., Ensslin, L., Ripoll-Feliu, V., & Dutra, A. (2015). Reflexões para avanço
    na área de Avaliação e Gestão do Desempenho das Universidades: Uma análise da literatura
    científica, Seminários em Administração (XVIII SEMEAD). São Paulo-SP.
CEC. (2005). Successful practices of environmental management systems in small and
    medium-size enterprises commission for environmental cooperation: A North American
    perspective. Commission for Environmental Cooperation.
Chen, L., & Meng, B. (2010). The application of value stream mapping based lean production
    system. *International Journal of Business and Management, 5*(6), 203–209.

Dombrowski, U., Crespo, I., & Zahn, T. (2010). Adaptive configuration of a lean production system in small and medium-sized enterprises. *Production Engineering, 4*(4), 341–348.

Dombrowski, U., & Mielke, T. (2014). Lean leadership–15 rules for a sustainable lean implementation. *Procedia CIRP, 17*, 565–570.

Doolen, T., & Hacker, M. (2005). A review of lean assessment in organizations: An exploratory study of lean practices by electronics manufacturers. *Journal of Manufacturing systems, 24*(1), 55–67.

Dora, M., Kumar, M., Van Goubergen, D., Molnar, A., & Gellynck, X. (2013). Operational performance and critical success factors of lean manufacturing in European food processing SMEs. *Trends in Food Science & Technology, 31*(2), 156–164.

Dora, M., Van Goubergen, D., Kumar, M., Molnar, A., & Gellynck, X. (2014). Application of lean practices in small and medium-sized food enterprises. *British Food Journal, 116*(1), 125–141.

Dutra, A., Ripoll-Feliu, V., Ensslin, S., & Ensslin, L. (2015). The construction of knowledge from the scientific literature about the theme seaport performance evaluation. *International Journal of Productivity and Performance Management, 64*(2), 243–269.

Ensslin, L., Ensslin, S., & Pinto, H. (2013). Processo de investigação e Análise bibliométrica: Avaliação da Qualidade dos Serviços Bancários. *Revista de Administração Contemporânea, 17* (3), 325–349.

Ensslin, S., Ensslin, L., Imlau, J., & Chaves, L. (2014). Processo de mapeamento das publicações científicas de um tema: Portfólio bibliográfico e análise bibliométrica sobre avaliação de desempenho de cooperativas de produção agropecuária. *Revista de Economia e Sociologia Rural, 52*(3), 587–608.

Ensslin, S., Ensslin, L., Matos, L., Dutra, A., & Ripoll-Feliu, V. (2015). Research opportunities in performance measurement in public utilities regulation. *International Journal of Productivity and Performance Management, 64*(7), 994–1017.

Godinho Filho, M., Ganga, G., & Gunasekaran, A. (2016). Lean manufacturing in Brazilian small and medium enterprises: Implementation and effect on performance. *International Journal of Production Research, 54*(24), 7523–7545.

Holland, C. R., & Light, B. (1999). A critical success factors model for ERP implementation. *Focus, IEEE Software*, 30–35.

Jadhav, J., Mantha, S., & Rane, S. (2014). Exploring barriers in lean implementation. *International Journal of Lean Six Sigma, 5*(2), 122–148.

Kumar, M., Antony, J., & Douglas, A. (2009). Does size matter for six sigma implementation? Findings from the survey in UK SMEs. *The TQM Journal, 21*(6), 623–635.

Kumar, M., Khurshid, K., & Waddell, D. (2014). Status of quality management practices in manufacturing SMEs: A comparative study between Australia and the UK. *International Journal of Production Research, 52*(21), 6482–6495.

Lacerda, R., Ensslin, L., & Ensslin, S. (2012). Uma análise bibliométrica da literatura sobre estratégia e avaliação de desempenho. *Gestão & Produção, 19*(1), 59–78.

Leone, N. (1991). A dimensão física das pequenas e médias empresas (PM E's): à procura de um critério homogeneizador. *Revista de Administração de Empresas, 31*(2), 53–59.

Marodin, G., & Saurin, T. (2015a). Classification and relationships between risks that affect lean production implementation: A study in Southern Brazil. *Journal of Manufacturing Technology Management, 26*(1), 57–79.

Marodin, G., & Saurin, T. (2015b). Managing barriers to lean production implementation: Context matters. *International Journal of Production Research, 53*(13), 3947–3962.

Matt, D. (2008). Template based production system design. *Journal of Manufacturing Technology Management, 19*(7), 783–797.

Matt, D., & Rauch, E. (2013). Implementation of lean production in small sized enterprises. *Procedia CIRP, 12*, 420–425.

Netland, T. (2016). Critical success factors for implementing lean production: The effect of contingencies. *International Journal of Production Research, 54*(8), 2433–2448.

Nordin, N., Deros, B., & Wahab, D. (2010). A survey on lean manufacturing implementation in Malaysian automotive industry. *International Journal of Innovation, Management and Technology, 1*(4), 374.

Ohno, T. (1998). *Toyota production system: Beyond large-scale production*. Portland: Productivity Press.

Paré, G., Trudel, M., Jaana, M., & Kitsiou, S. (2015). Synthesizing information systems knowledge: A typology of literature reviews. *Information & Management, 52*(2), 183–199.

Rose, A., Deros, B., & Ab Rahman, M. (2013). A study on lean manufacturing implementation in Malaysian automotive component industry. *International Journal of Automotive and Mechanical Engineering, 8,* 1467.

Saurin, T., Ribeiro, J., & Marodin, G. (2010). Identificação de oportunidades de pesquisa a partir de um levantamento da implantação da produção enxuta em empresas do Brasil e do exterior. *Gestão & Produção, 17*(4), 829–841.

SEBRAE. (2013). Sobrevivência das Empresas no Brasil: Coleção estudos e pesquisas, 69. Brasília.

SEBRAE. (2014). Participação das Micro e Pequenas Empresas na Economia Brasileira, 106. Brasília.

Shah, R., & Ward, P. (2003). Lean manufacturing: Context, practice bundles, and performance. *Journal of operations management, 21*(2), 129–149.

Timans, W., Antony, J., Ahaus, K., & van Solingen, R. (2012). Implementation of lean six sigma in small-and medium-sized manufacturing enterprises in the Netherlands. *Journal of the Operational Research Society, 63*(3), 339–353.

Tortorella, G., Marodin, G., Fogliatto, F., & Miorando, R. (2015a). Learning organisation and human resources management practices: An exploratory research in medium-sized enterprises undergoing a lean implementation. *International Journal of Production Research, 53*(13), 3989–4000.

Tortorella, G., Marodin, G., Miorando, R., & Seidel, A. (2015b). The impact of contextual variables on learning organization in firms that are implementing lean: A study in Southern Brazil. *The International Journal of Advanced Manufacturing Technology, 78*(9–12), 1879–1892.

Treville, S., & Antonakis, J. (2006). Could lean production job design be intrinsically motivating? Contextual, configurational, and levels-of-analysis issues. *Journal of Operations Management, 24*(2), 99–123.

Womack, J., & Jones, D. (1992). *A máquina que mudou o mundo*. Rio de Janeiro: Campus.

Wong, Y., Wong, K., & Ali, A. (2009). A study on lean manufacturing implementation in the Malaysian electrical and electronics industry. *European Journal of Scientific Research, 38*(4), 521–535.

Worley, J., & Doolen, T. (2006). The role of communication and management support in a lean manufacturing implementation. *Management Decision, 44*(2), 228–245.

Yew Wong, K. (2005). Critical success factors for implementing knowledge management in small and medium enterprises. *Industrial Management & Data Systems, 105*(3), 261–279.

Yang, P., & Yu, Y. (2010). The Barriers to SMEs' implementation of lean production and counter measures—based on SMS in Wenzhou. *International Journal of Innovation, Management and Technology, 1*(2), 220–225.

Zhou, B. (2016). Lean principles, practices, and impacts: A study on small and medium-sized enterprises (SMEs). *Annals of Operations Research, 241*(1–2), 457–474.

Appendix—Bibliographical Portfolio

Abolhassani, A., Layfield, K., & Gopalakrishnan, B. (2016). Lean and US manufacturing industry: Popularity of practices and implementation barriers. *International Journal of Productivity and Performance Management, 65*(7), 875–897.

Achanga, P., Shehab, E., Roy, R., & Nelder, G. (2006). Critical success factors for lean implementation within SMEs. *Journal of Manufacturing Technology Management, 17*(4), 460–471.

Anand, G., & Kodali, R. (2010). Analysis of lean manufacturing frameworks. *Journal of Advanced Manufacturing Systems, 9*(1), 1–30.

Antony, J., Kumar, M., & Madu, C. N. (2005). Six sigma in small-and medium-sized UK manufacturing enterprises: Some empirical observations. *International Journal of Quality & Reliability Management, 22*(8), 860–874.

Bakas, O., Govaert, T., & Van Landeghem, H. (2011). Challenges and success factors for implementation of lean manufacturing in European SMES. In *13th International conference on the Modern Information Technology in the Innovation Processes of the Industrial Enterprise (MITIP 2011)* (Vol. 1). Tapir Academic Press.

Belhadi, A., & Touriki, F. (2016). A framework for effective implementation of lean production in small and medium-sized enterprises. *Journal of Industrial Engineering and Management, 9*(3), 786–810.

Bhamu, J., & Singh Sangwan, K. (2014). Lean manufacturing: Literature review and research issues. *International Journal of Operations & Production Management, 34*(7), 876–940.

Bhasin, S. (2012). An appropriate change strategy for lean success. *Management Decision, 50*(3), 439–458.

Bhasin, S. (2012). Prominent obstacles to lean. *International Journal of Productivity and Performance Management, 61*(4), 403–425.

Dombrowski, U., Crespo, I., & Zahn, T. (2010). Adaptive configuration of a lean production system in small and medium-sized enterprises. *Production Engineering, 4*(4), 341–348.

Dombrowski, U., & Mielke, T. (2014). Lean leadership–15 rules for a sustainable lean implementation. *Procedia CIRP, 17,* 565–570.

Doolen, T., & Hacker, M. (2005). A review of lean assessment in organizations: An exploratory study of lean practices by electronics manufacturers. *Journal of Manufacturing Systems, 24*(1), 55–67.

Dora, M., Kumar, M., Van Goubergen, D., Molnar, A., & Gellynck, X. (2013). Operational performance and critical success factors of lean manufacturing in European food processing SMEs. *Trends in Food Science & Technology, 31*(2), 156–164.

Dora, M., Van Goubergen, D., Kumar, M., Molnar, A., & Gellynck, X. (2014). Application of lean practices in small and medium-sized food enterprises. *British Food Journal, 116*(1), 125–141.

Godinho Filho, M., Ganga, G., & Gunasekaran, A. (2016). Lean manufacturing in Brazilian small and medium enterprises: Implementation and effect on performance. *International Journal of Production Research, 54*(24), 7523–7545.

Hallgren, M., & Olhager, J. (2009). Lean and agile manufacturing: External and internal drivers and performance outcomes. *International Journal of Operations & Production Management, 29*(10), 976–999.

Jadhav, J., Mantha, S., & Rane, S. (2014). Exploring barriers in lean implementation. *International Journal of Lean Six Sigma, 5*(2), 122–148.

Kumar, M., & Antony, J. (2009). Multiple case-study analysis of quality management practices within UK six sigma and non-six sigma manufacturing small-and medium-sized enterprises. *Proceedings of the Institution of Mechanical Engineers, Part B: Journal of Engineering Manufacture, 223*(7), 925–934.

Kumar, M., Antony, J., & Douglas, A. (2009). Does size matter for six sigma implementation? Findings from the survey in UK SMEs. *The TQM Journal, 21*(6), 623–635.

Kumar, M., Khurshid, K., & Waddell, D. (2014). Status of quality management practices in manufacturing SMEs: A comparative study between Australia and the UK. *International Journal of Production Research, 52*(21), 6482–6495.

Kumar, M., Antony, J., Singh, R., Tiwari, M., & Perry, D. (2006). Implementing the Lean Sigma framework in an Indian SME: A case study. *Production Planning and Control, 17*(4), 407–423.

Manville, G., Greatbanks, R., Krishnasamy, R., & Parker, D. W. (2012). Critical success factors for lean six sigma programmes: A view from middle management. *International Journal of Quality & Reliability Management, 29*(1), 7–20.

Marodin, G., & Saurin, T. (2015a). Classification and relationships between risks that affect lean production implementation: A study in Southern Brazil. *Journal of Manufacturing Technology Management, 26*(1), 57–79.

Marodin, G., & Saurin, T. (2015b). Managing barriers to lean production implementation: Context matters. *International Journal of Production Research, 53*(13), 3947–3962.

Matt, D., & Rauch, E. (2013). Implementation of lean production in small sized enterprises. *Procedia CIRP, 12,* 420–425.

Netland, T. (2016). Critical success factors for implementing lean production: The effect of contingencies. *International Journal of Production Research, 54*(8), 2433–2448.

Nordin, N., Deros, B., Wahab, D., & Rahman, M. (2012). A framework for organisational change management in lean manufacturing implementation. *International Journal of Services and Operations Management, 12*(1), 101–117.

Nordin, N., Deros, B., & Wahab, D. (2010). A survey on lean manufacturing implementation in Malaysian automotive industry. *International Journal of Innovation, Management and Technology, 1*(4), 374.

Rose, A., Deros, B., & Ab Rahman, M. (2013). A study on lean manufacturing implementation in Malaysian automotive component industry. *International Journal of Automotive and Mechanical Engineering, 8,* 1467.

Rose, A., Deros, B., Rahman, M., & Nordin, N. (2011). Lean manufacturing best practices in SMEs, In *Proceedings of the 2011 International Conference on Industrial Engineering and Operations Management* (Vol. 2, No. 5, pp. 872–877).

Sánchez, M., & Pérez, M. (2001). Lean indicators and manufacturing strategies. *International Journal of Operations & Production Management, 21*(11), 1433–1452.

Saurin, T., Ribeiro, J., & Marodin, G. (2010). Identificação de oportunidades de pesquisa a partir de um levantamento da implantação da produção enxuta em empresas do Brasil e do exterior. *Gestão e Produção, 17*(4), 829–841.

Shah, R., & Ward, P. (2007). Defining and developing measures of lean production. *Journal of Operations Management, 25*(4), 785–805.

Shah, R., & Ward, P. (2003). Lean manufacturing: Context, practice bundles, and performance. *Journal of Operations Management, 21*(2), 129–149.

Sim, K., & Rogers, J. (2008). Implementing lean production systems: Barriers to change. *Management Research News, 32*(1), 37–49.

Timans, W., Antony, J., Ahaus, K., & van Solingen, R. (2012). Implementation of lean six sigma in small-and medium-sized manufacturing enterprises in the Netherlands. *Journal of the Operational Research Society, 63*(3), 339–353.

Tortorella, G., Marodin, G., Fogliatto, F., & Miorando, R. (2015a). Learning organisation and human resources management practices: An exploratory research in medium-sized enterprises undergoing a lean implementation. *International Journal of Production Research, 53*(13), 3989–4000.

Tortorella, G., Marodin, G., Miorando, R., & Seidel, A. (2015b). The impact of contextual variables on learning organization in firms that are implementing lean: A study in Southern Brazil. *The International Journal of Advanced Manufacturing Technology, 78*(9–12), 1879–1892.

Wilson, M., & Roy, R. (2009). Enabling lean procurement: A consolidation model for small-and medium-sized enterprises. *Journal of Manufacturing Technology Management, 20*(6), 817–833.

Wong, Y., Wong, K., & Ali, A. (2009). A study on lean manufacturing implementation in the Malaysian electrical and electronics industry. *European Journal of Scientific Research, 38*(4), 521–535.

Worley, J., & Doolen, T. (2006). The role of communication and management support in a lean manufacturing implementation. *Management Decision, 44*(2), 228–245.

Yew Wong, K. (2005). Critical success factors for implementing knowledge management in small and medium enterprises. *Industrial Management & Data Systems, 105*(3), 261–279.

Yang, P., & Yu, Y. (2010). The Barriers to SMEs' implementation of lean production and counter measures—based on SMS in Wenzhou. *International Journal of Innovation, Management and Technology, 1*(2), 220–225.

Zhou, B. (2016). Lean principles, practices, and impacts: A study on small and medium-sized enterprises (SMEs). *Annals of Operations Research, 241*(1–2), 457–474.

Application of Structural Equation Modeling for Analysis of Lean Concepts Deployment in Healthcare Sector

S. Vinodh and A. M. Dhakshinamoorthy

Abstract Healthcare sector had been recently witnessing lean concepts deployment. In order to enable effective implementation of lean concept in the healthcare domain, a structural model needs to be developed. The goal of this study is to analyze the relationship between lean constructs and healthcare performance. The model had been developed with five constructs and 20 measurement variables. The model was developed through the literature on lean deployment in healthcare unit and responses from healthcare experts. Forty-five responses were collected from healthcare experts. Partial least square (PLS) based Structural equation modeling (SEM) approach had been used for analysis. Based on the simulation results, it was found that "management" is the most driving enabler which influences people, process, technology, and other resources of the healthcare unit. The reliability of the developed model has been tested using Cronbach's alpha and composite reliability. The study presents an attempt to develop a statistical model for lean concepts deployment so as to enhance healthcare performance.

Keywords Lean · Leanness · Statistical modeling · Structural equation modeling (SEM) · Healthcare sector

1 Introduction

Lean concepts were initially applied in the manufacturing sector. During recent years, service sectors have started to adopt lean practices to reduce service cost. But, lean applications in healthcare sector have not been fully explored. Healthcare practices are subjected to developments in terms of increased number of patients,

S. Vinodh (✉) · A. M. Dhakshinamoorthy
Department of Production Engineering, National Institute of Technology,
Tiruchirappalli 620 015, Tamil Nadu, India
e-mail: vinodh_sekar82@yahoo.com

A. M. Dhakshinamoorthy
e-mail: dhakshinamoorthy92@gmail.com

© Springer International Publishing AG, part of Springer Nature 2018
J. P. Davim (ed.), *Progress in Lean Manufacturing*,
Management and Industrial Engineering,
https://doi.org/10.1007/978-3-319-73648-8_4

lower waiting time, and cost-effective procedures. The cost of healthcare procedures is increasing at a rapid rate (Vegting et al. 2012; Reijula and Tommelein 2012; Shazali et al. 2013). There exists a need to reduce costs and increase their efficiency, by providing an improved service with reduced service cost to the customer. Adoption of lean concepts in health care is one of the prime ways to achieve the need for improved service level. In order to enable effective deployment of lean in healthcare applications, a statistical model needs to be developed depicting the interrelationship between the constructs. The research objective of the study reported in this chapter is to identify appropriate lean constructs and to develop a statistical model linking the relation between lean concepts implementation and performance improvements. In this context, this chapter presents a study in which five constructs and 20 indicators are being developed based on literature review. Based on expert inputs, the model has been developed and analyzed using Structural equation modeling approach. The practical inferences are being derived. This chapter is organized as follows: Introduction section is followed by literature review, methodology, SEM model development, Inferences from SEM model, and conclusions.

2 Literature Review

The literature has been reviewed from the perspectives of Lean concepts deployment in healthcare domain and studies on modeling with reference to the healthcare sector.

2.1 Review on Lean Concepts Deployment in the Healthcare Sector

Dahlgaard et al. (2011) developed a conceptual model to assess, measure, diagnose, and improve healthcare organizations. Five universities and five hospitals in Denmark were considered in the study. Based on the response from employees and management, ILL index (Innovativeness, Learning, and Lean) was calculated to measure the level of excellence prevailing in healthcare units.

de Souza and Pidd (2011) studied the similarities and differences in implementing lean concepts in the manufacturing sector and healthcare sector. They discussed the factors that hinder the successful implementation of lean practices in the healthcare sectors. Some of the barriers they have shown are resistance to change, management hierarchy, professional skills, and performance measurement methods.

Radnor et al. (2012) performed four multilevel case studies on lean implementation in the English National Health Service (NHS). It was observed that the healthcare sectors lacked in embracing broad lean thinking for system-wide bene-fits. Gitlow et al. (2013) analyzed the factors that contribute to preventing medical accidents in healthcare units. A model was developed to suggest "Standard Best Practices" in healthcare units.

Costa and Godinho Filho (2016) carried out a literature-based study. One hundred and seven research papers in lean healthcare were evaluated and recent trends were identified. In the study, lean implementation in health care in different countries was analyzed. It was observed that 5S and Value Stream Mapping (VSM) were the most frequently used lean tools in the healthcare sector.

Hussain et al. (2016) conducted a study to assist the deployment of lean in healthcare delivery system. In the study, Analytical Hierarchical Process (AHP) was used to decompose complex, unstructured issues prevailing in lean healthcare and organize them into a set of components in a multilevel hierarchical form. Twenty-one healthcare wastes were identified and ranked based on the responses from experienced healthcare professionals.

Setijono et al. (2010) conducted a study in Swedish emergency ward to find the best number of surgeons and doctors to be employed so as to reduce patients' Non-Value Added Time (NVAT) and Total Time in the System (TTS). An ARENA model of the patient service cycle was developed and simulated to infer the results. Thirteen percent reduction of patients' NVAT was achieved in the study.

Jorma et al. (2016) studied the benefits of lean concept implemented in healthcare for managing the treatment process. They have done a questionnaire survey with 248 responders working in the healthcare sector for analyzing the response regarding the implementation of lean in health care. From the survey, they got positive responses regarding benefits of successful implementation of lean tools and techniques.

Lindskog et al. (2016) aimed toward finding the extent of lean practices implementation in health care. Multiple linear regression models have been used to study the extent of implementation of lean tools. They concluded that VSM, 5S, and standardized work lean tools that can enhance the working condition of employees in the healthcare sector.

Kovacevic et al. (2016) reviewed the lean tools and techniques that have been successfully implemented in the healthcare sectors. They reviewed the benefits of most frequently used lean tools, namely 5S, VSM, Kaizen, and Visual management. They found several benefits of implementing lean tools. They are reduction in patient average waiting time, significant improvement in flow time, and reduction in manpower.

Eriksson et al. (2016) conducted semi-structured interviews in two Swedish hospitals to understand practitioners perception on lean deployment. It was observed that lean implementation being a bottom-up approach often reflected in conflicting ideas with top management which usually follows top-down approach. The study emphasized the necessity to frame lean practices that have consensus from both technical experts and top management.

2.2 Review of Modeling Studies in Healthcare Sector

Hussey and Eagan (2007) used SEM methodology to validate the developed environmental performance model on environmental management systems in Small and Medium Enterprises (SMEs). From the results, it was observed that "leadership" greatly influences "environmental planning" and "environmental management" which in turn greatly influences "environmental results".

Hussey and Eagan (2007) developed a model to understand the factors influencing customer satisfaction in the healthcare sector. "High performance work system", "service quality", "customer orientation" were identified as the critical factors. The model was statistically validated using SEM approach. It was observed that "High performance work system" acted as the key driving factor which significantly influenced "customer orientation" and "service quality". It was also observed that all factors significantly influenced in providing better customer satisfaction.

Martín-Consuegra et al. (2007) analyzed the interaction among price fairness, customer satisfaction, loyalty, and their overall impact on price acceptance by the customers. A service industry in the airline sector was identified as the case unit to collect the responses. Based on the responses, a path model was developed which indicated the factor relationship that influenced price acceptance by the customers. SEM approach was used to validate the model. It was observed that price fairness greatly influenced customer satisfaction and loyalty; customer satisfaction has significant influence on loyalty; price fairness, customer satisfaction, and loyalty had combined significant influence on price acceptance.

Chahal and Bala (2012) studied the interrelations among factors influencing service brand equity in the healthcare sector. SEM approach was used to analyze the data collected from 206 respondents. It was observed that brand loyalty and perceived quality had a significant influence on brand equity. Brand image was found to be mediating variable which had an indirect effect on brand equity through brand loyalty.

Hussain et al. (2016) discussed the successful lean practices implemented in the healthcare sector. They have shown a case study of a public healthcare system situated in Abu Dhabi. They also discussed seven wastes associated with lean practices in health care along with 21 sub-waste. Their aim was to prioritize the influential waste with the help of Analytical Hierarchy Process (AHP) and selection of waste minimization technique. They found that transportation waste and inventory waste are the most critical wastes pertaining to the healthcare sector.

Tarhini et al. (2017) analyzed the effect of cultural values at the individual level on user's acceptance of e-learning. Responses were collected from 569 undergraduate and postgraduate students regarding their usage of e-learning tools. A statistical model was developed and validated using SEM approach. "Quality of work life", "Uncertainty avoidance", "Subjective norms", and "Behavioral intention" were the four cultural dimensions considered in the study. It was observed that all four constructs had a significant influence on "Technology acceptance level".

Patri and suresh (2017) aimed toward finding the relationship between factors influencing implementation of lean practices in health care. They made a hierarchy structure of considered factors using Interpretive Structural Modeling. Ranking of factors was done based on the driving power and dependence power of factors. From the analysis, they found that leadership is the most important factor in successful implementation of lean practices in the healthcare sector.

Mitchell et al. (2017) aimed toward finding the relationship between size of the hospital, patient comorbidity and time to customer in hospital, i.e., time difference between admission and discharge with the help of SEM. They concluded that patient comorbidity had a great influence on timing of infection.

Zhao et al. (2017) studied the direct and indirect effects of socioeconomic status on rectal cancer risk in patients. Dietary patterns were collected among 39 food groups from the people of Newfoundland and Labrador. SEM approach was used in the study and the results showed that dietary pattern greatly influenced socioeconomic status among people which in turn greatly influenced the risk of acquiring rectal cancer.

2.3   Research Gaps

Though there are certain studies reported on lean concepts adoption in the healthcare domain, and statistical modeling for lean concepts adoption is attempted, there exists a potential for developing a statistical model linking lean concepts adoption and operational performance.

3   Methodology

The methodological steps are shown in Fig. 1. Critical parameters that influence lean deployment in the healthcare sector were identified as constructs. Based on the literature review and inputs from healthcare practitioners, measured variables corresponding to each construct were developed and a model depicting the interrelations between constructs was being developed. The model developed entails five constructs and 20 measured variables. A case healthcare unit in the Tamil Nadu state of India is identified for the study. Forty-five responses were collected from healthcare practitioners of the case healthcare unit for the developed SEM model. The model is then tested for "reliability" and the results are obtained by simulating the model.

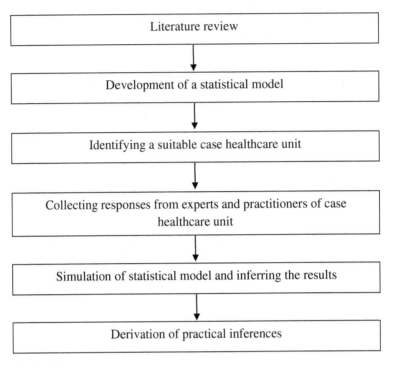

Fig. 1 Methodology

4 Development of Statistical Model Using SEM

Structural Equation Modeling (SEM) is a multivariate statistical approach that is used to analyze structural relationships. It combines factor analysis and multiple regression analysis for analysis of the structural relationship between measured variables and latent constructs. Multiple relationships among factors present in the model and their combined interaction effect can be simultaneously estimated using SEM in a single analysis.

4.1 PLS-SEM

Partial Least Squares (PLS) approach to SEM is highly beneficial for basic investigation. It is an alternative to covariance-based SEM. It is used for problems involving formative indicators, small sample size, and non-normal data. It includes two sub-models: measurement and structural model. Measurement model depicts relationships between observed data and latent variables; whereas structural model indicates relationships between latent variables. PLS-SEM is used to explore, analyze, and test the conceptual models and theory.

4.2 Cronbach's Alpha

Cronbach's alpha is a measure of internal consistency, i.e., the measurement of closeness between set of items belonging to a group. If alpha value is high, it implies that the measure is non-unidimensional. It is a measure of scale reliability. Apart from internal consistency measurement, it is desired to provide evidence that the subject scale is unidimensional, further analyses can be performed. Based on research studies, it was found that the value of Cronbach's alpha of 0.7 or higher is preferred.

4.3 Composite Reliability

The composite reliability is typically parameter examined when PLS is applied as analysis method. It is the measure of internal consistency. It is a measure of overall reliability of a set of heterogeneous but similar items.

In SEM model, the constructs used can be classified into two types namely: endogenous and exogenous constructs. Endogenous constructs are equivalent to dependent variables and exogenous constructs are equivalent to independent variables used in multiple regression model.

The following steps are involved in building the model:

Step 1: Defining individual construct—proper definition has to be given for each construct theoretically.

Step 2: Developing the measurement model—it is made to show the relationship between exogenous variables and endogenous variable, here an arrow has to be drawn from measured variable to construct.

Step 3: Design the model to provide empirical results: here, the model has to be designed by minimizing the identification problem. This can be done by using rank condition and order condition method.

Step 4: Specify the structure model: Here, the structural path has to be drawn in between constructs. In this model, an arrow cannot enter into the exogenous construct. For representing the structural relationship between any two constructs, a single-headed arrow has to be used.

Step 5: Examine the structural model validity—Chi-Square test has been used to validate the structural model.

The model developed entails five constructs and 20 measured variables. The constructs and variables are developed through the literature on lean deployment in health care and responses from healthcare experts and academicians. Table 1 shows the constructs and their corresponding measured variables being identified. Constructs are being identified as Management people, process, technology and lean performance. For example, Management Construct consists of Variables such as Top management support, Organizational structure and Incentives and training program.

Table 1 Constructs and measured variables identified for the SEM model

S. No.	Construct	Measured variables
1	*Management* Mazzocato et al. (2010)	Top management support (M1)
		Organizational structure (M2)
		Incentives and training program (M3)
2	*People* Dahlgaard et al. (2011) Dellve et al. (2015)	Staff involvement (P1)
		Organizational culture (P2)
		Change management (P3)
3	*Process* Doolen and Hacker (2005) Bhasin (2011) Wahab et al. (2013)	Housekeeping (Pr1)
		Standard operating procedures (Pr2)
		Benchmarking (Pr3)
		Visual management (Pr4)
		Operational environment (Pr5)
4	*Technology* Vinodh and Chintha (2011) Sharma et al. (2015)	Time management (T1)
		Status of quality (T2)
		Cost management (T3)
5	*Lean performance* Dahlgaard et al. (2011) Radnor et al. (2012) Costa and Godinho Filho (2016)	Streamlined process (LP1)
		Quality of service (LP2)
		Reduced patient stay (LP3)
		Patient satisfaction (LP4)
		Staff morale (LP5)
		Cost incurred (LP6)

Forty-five responses were collected from healthcare experts in the case healthcare unit. Structure of the model developed, results of the model's composite reliability, Cronbach's alpha values of the constructs pertaining to the SEM model and inferential results from model simulation are shown in Figs. 2, 3, 4 and 5 respectively. Figure 2 shows the relationship between the constructs and its associated latent variables. The model consists of five constructs namely People, Management, Process, Technology and Lean Performance. Indicator reliability and internal consistency reliability are the two main reliability constraints that are to be checked. The reliability values are computed in SmartPLS by using the command Calculate ≫ PLS algorithm. Indicator reliability is computed as cronbach's alpha value and internal consistency reliability is computed as composite reliability value. The indicator reliability value and internal consistency reliability value must be greater than 0.7 for accepting the construct. If the values are found to be less than 0.7, further scale refinement has to done by either increasing the sample size or by altering the input values. Figures 3 and 4 show the obtained composite reliability values and Cronbach's alpha values for the developed structural model. Figure 5 shows the cumulative path values of the latent variables based on the significance of relationships among the constructs.

Composite reliability test and Cronbach's alpha test are conducted to find the reliability of the developed SEM model. From the simulation results, it is observed

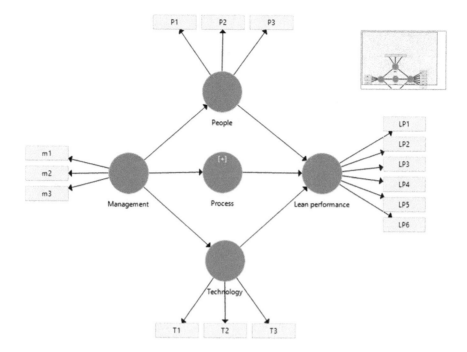

Fig. 2 Snapshot depicting relationship among constructs

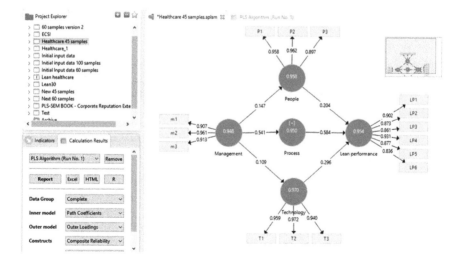

Fig. 3 Composite reliability of the developed SEM model

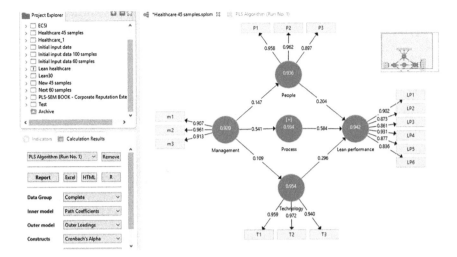

Fig. 4 Cronbach's alpha values of the developed SEM model

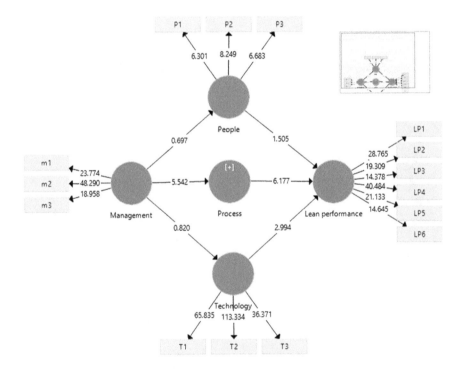

Fig. 5 SEM results for depicting the significance of relationships among constructs

that "management" is the most driving enabler which significantly influences "people", "process", "technology" and other resources present in the healthcare unit. It is also observed that "people", "process", and "technology" individually have significant influence in deploying lean practices in the healthcare sector.

5 Practical Implications

The developed model enabled the medical practitioners and experts to systematically develop the statistical model linking the identified Constructs and operational performance. The study also enabled the identification of most driving enablers for lean concepts adoption in healthcare domain.

6 Conclusions

The main principle of lean is to eliminate waste activities which can be carried out by identifying critical and bottleneck stations in the process flow and bringing remedial measures so as to achieve a seamless flow for the overall process. From a healthcare perspective, lean practices can reduce patient waiting time, reduce patient stay in healthcare unit, increase visibility of the activities involved in health care, reduces medical defects, increases morale of staffs and practitioners in healthcare units, and reduces the cost involved in providing the service without compromising the quality of service being delivered. Unlike in manufacturing sector, lean practices which initially had its inception in shop floor cannot be directly adopted to service sector like healthcare services. In the present study, the critical factors influencing lean deployment in the healthcare sector is identified and their effect on establishing lean in health care is validated using SEM approach. It is observed that all four constructs namely, "management", "people", "process", and "technology" have a significant influence on the successful deployment of lean in the healthcare sector. It is also observed that "management" being the prime driving factor has significant contribution in enhancing "people", "process", "technology" constructs, and other resources used in providing the healthcare service.

7 Limitations and Future Work

The study is focused on a single healthcare unit. In future, more number of studies could be conducted with reference to various other healthcare units to improve the practical validity of the approach. In the present study, 20 factors are considered. In future, number of factors could be increased to deal with managerial advancements in healthcare domain. Also, a structural model could be developed based on ISM approach and the model could further be statistically validated using SEM.

References

Bhasin, S. (2011). Measuring the Leanness of an organisation. *International Journal of Lean Six Sigma, 2*(1), 55–74.

Chahal, H., & Bala, M. (2012). Significant components of service brand equity in healthcare sector. *International Journal of Health Care Quality Assurance, 25*(4), 343–362.

Costa, L. B. M., & Godinho Filho, M., (2016). Lean healthcare: Review, classification and analysis of literature. *Production Planning & Control, 27*(10), 823–836.

Dahlgaard, J. J., Pettersen, J., & Dahlgaard-Park, S. M. (2011). Quality and lean health care: A system for assessing and improving the health of healthcare organisations. *Total Quality Management & Business Excellence, 22*(6), 673–689.

Dellve, L., Williamsson, A., Strömgren, M., Holden, R. J., & Eriksson, A. (2015). Lean implementation at different levels in Swedish hospitals: The importance for working conditions and stress. *International Journal of Human Factors and Ergonomics, 3*(3–4), 235–253.

de Souza, L. B., & Pidd, M. (2011). Exploring the barriers to lean health care implementation. *Public Money & Management, 31*(1), 59–66.

Doolen, T. L., & Hacker, M. E. (2005). A review of lean assessment in organizations: An exploratory study of lean practices by electronics manufacturers. *Journal of Manufacturing systems, 24*(1), 55–67.

Eriksson, N., Müllern, T., Andersson, T., Gadolin, C., Tengblad, S., & Ujvari, S. (2016). Involvement drivers: A study of nurses and physicians in improvement work. *Quality Management in Healthcare, 25*(2), 85–91.

Gitlow, H., "Amy" Zuo, Q., Ullmann, S. G., Zambrana, D., Campo, R. E., Lubarsky, D., et al. (2013). The causes of never events in hospitals. *International Journal of Lean Six Sigma, 4*(3), 338–344.

Hussain, M., Malik, M., & Al Neyadi, H. S. (2016). AHP framework to assist lean deployment in Abu Dhabi public healthcare delivery system. *Business Process Management Journal, 22*(3), 546–565.

Hussey, D. M., & Eagan, P. D. (2007). Using structural equation modeling to test environmental performance in small and medium-sized manufacturers: Can SEM help SMEs? *Journal of Cleaner Production, 15*(4), 303–312.

Jorma, T., Tiirinki, H., Bloigu, R., & Turkki, L. (2016). Lean thinking in Finnish healthcare. *Leadership in Health Services, 29*(1), 9–36.

Kovacevic, M., Jovicic, M., Djapan, M., & Zivanovic-Macuzic, I. (2016). Lean thinking in healthcare: Review of implementation results. *International Journal for Quality Research, 10* (1).

Lindskog, P., Hemphälä, J., Eklund, J., & Eriksson, A. (2016). Lean in healthcare: Engagement in development, job satisfaction or exhaustion? *Journal of Hospital Administration, 5*(5), 91.

Martín-Consuegra, D., Molina, A., & Esteban, Á. (2007). An integrated model of price, satisfaction and loyalty: An empirical analysis in the service sector. *Journal of Product & Brand Management, 16*(7), 459–468.

Mazzocato, P., Savage, C., Brommels, M., Aronsson, H., & Thor, J. (2010). Lean thinking in healthcare: A realist review of the literature. *Quality and Safety in Health Care, 19*(5), 376–382.

Mitchell, B. G., Anderson, M., & Ferguson, J. K. (2017). A predictive model of days from infection to discharge in patients with healthcare-associated urinary tract infections: A structural equation modelling approach. *Journal of Hospital Infection, 97*(3), 282–287.

Patri, R., & Suresh, M. (2017). Factors influencing lean implementation in healthcare organizations: An ISM approach. *International Journal of Healthcare Management*, 1–13. https://doi.org/10.1080/20479700.2017.1300380.

Radnor, Z. J., Holweg, M., & Waring, J. (2012). Lean in healthcare: The unfilled promise? *Social Science and Medicine, 74*(3), 364–371.

Reijula, J., & Tommelein, I. D. (2012). Lean hospitals: A new challenge for facility designers. *Intelligent Buildings International, 4*(2), 126–143.

Setijono, D., Mohajeri Naraghi, A., & Pavan Ravipati, U. (2010). Decision support system and the adoption of lean in a swedish emergency ward: Balancing supply and demand towards improved value stream. *International Journal of Lean Six Sigma, 1*(3), 234–248.

Sharma, V., Dixit, A. R., & Qadri, M. A. (2015). Impact of lean practices on performance measures in context to Indian machine tool industry. *Journal of Manufacturing Technology Management, 26*(8), 1218–1242.

Shazali, N. A., Habidin, N. F., Ali, N., Khaidir, N. A., & Jamaludin, N. H. (2013). Lean healthcare practice and healthcare performance in Malaysian healthcare industry. *International Journal of Scientific and Research Publications, 3*(1), 1–5.

Tarhini, A., Hone, K., Liu, X., & Tarhini, T. (2017). Examining the moderating effect of individual-level cultural values on users' acceptance of E-learning in developing countries: A structural equation modeling of an extended technology acceptance model. *Interactive Learning Environments, 25*(3), 306–328.

Vegting, I. L., van Beneden, M., Kramer, M. H. H., Thijs, A., Kostense, P. J., & Nanayakkara, P. W. (2012). How to save costs by reducing unnecessary testing: Lean thinking in clinical practice. *European Journal of Internal Medicine, 23*(1), 70–75.

Vinodh, S., & Chintha, S. K. (2011). Leanness assessment using multi-grade fuzzy approach. *International Journal of Production Research, 49*(2), 431–445.

Vinodh, S., & Joy, D. (2012). Structural equation modelling of lean manufacturing practices. *International Journal of Production Research, 50*(6), 1598–1607.

Wahab, A. N. A., Mukhtar, M., & Sulaiman, R. (2013). A conceptual model of lean manufacturing dimensions.*Procedia Technology, 11*, 1292–1298.

Zhao, J., Zhu, Y., & Wang, P. (2017). Examining the direct and indirect effects of socioeconomic status (SES) on colorectal cancer risk using structural equation modeling. *European Journal of Cancer, 72*, S56.

Association Between Lean Manufacturing Teaching Methods and Students' Learning Preferences

Guilherme Luz Tortorella, Rogério Miorando and Aurora Patricia Piñeres Castillo

Abstract As more companies embrace the concept of Lean Manufacturing (LM), universities should consider changing their curricula since there is a high likelihood that students will participate in some aspects of LM as they begin their professional careers. Thus, it is important to provide appropriate learning experiences to prepare students prior to LM. However, students may learn through several ways and teaching LM has proved to be extremely challenging, especially for engineering students who are not used to abstractions. This paper aims at examining the association between different LM teaching methods and students' learning preferences to increase their performance in courses. To achieve that, 76 graduate students from Industrial Engineering, who participated in two LM courses with different teaching methods were assessed according to their learning preferences and performance. Each LM course had a specific teaching approach: (i) classroom lectures and exercises (classified as traditional teaching methods), and (ii) problem-based learning (PBL) in real-world problems from companies undergoing an LM implementation. From the eight hypotheses formulated for this study, our results verified four of them, being two for each LM teaching method applied. Further, the effect of the learning dimension "information perception" seems to be more extensive than the others, since students' performance in both teaching methods is significantly associated with it. The mix between active learning methods and traditional teaching methods may facilitate dialogical learning, encouraging collaboration between students and facilitate the transfer of knowledge on LM.

G. L. Tortorella (✉) · R. Miorando
Universidade Federal de Santa Catarina, Florianópolis, Brazil
e-mail: gtortorella@bol.com.br

R. Miorando
e-mail: miorando@gmail.com

A. P. P. Castillo
Corporación Universidad de La Costa, Barranquilla, Colombia
e-mail: apineres2@cuc.edu.co

© Springer International Publishing AG, part of Springer Nature 2018
J. P. Davim (ed.), *Progress in Lean Manufacturing*,
Management and Industrial Engineering,
https://doi.org/10.1007/978-3-319-73648-8_5

105

Keywords Lean manufacturing · Teaching methods · Learning styles
Students' performance

1 Introduction

Lean manufacturing (LM) implementation entails fundamental changes in companies' managerial systems, across the organizational and department levels (Karlsson
and Ahlstom 1996; Wan and Chen 2008). LM implementation is about changing
both technical and sociocultural aspects (Tortorella and Fogliatto 2014), in order to
promote an organizational culture that systematically seeks for waste elimination
and quality improvement through people empowerment (Sawhney and Chason
2005; Womack and Jones 2009). Despite being originally conceived in manufacturing context, LM is becoming popular in other domains, such as healthcare,
administration, etc. Such diversified implementation and wide relevance, gives rise
to the issue of teaching this topic to students and practitioners from different fields
(Dukovska-Popovska et al. 2008). Hence, as change is introduced into companies
through LM implementation, universities should consider changing their curriculum with this movement (Stier 2003; Kanigolla et al. 2014), since there is a high
likelihood that students will encounter and participate in some aspects of LM as
they begin internships or full-time employment (Conger and Miller 2014;
Suárez-Barraza and Rodríguez-González 2015). In this sense, it is important that
universities provide appropriate learning experiences in the curriculum to prepare
students prior to LM (Kahlen et al. 2011).

However, students may learn through several ways, such as hearing, practicing,
visualizing, etc.; meanwhile, the teaching methods are also subject to variation. For
instance, some instructors prefer discussions or demonstrations, while others adopt
lectures (Kaliská 2012). In this sense, learning effectiveness is governed in part by
students' native ability and prior preparation, but also by the compatibility of their
learning preferences and the applied teaching methods (Felder and Silverman 1988;
Felder and Spurlin 2005). Mismatches between students' learning preferences and
professors' teaching methods entail bored and inattentive students who tend to
poorly perform on tests and, ultimately, change to other curricula (Litzinger et al.
2007). On the other hand, extreme approaches such as complete individualized
instruction or one-size-fits-all are either impractical or ineffective for most students
(Ramsden 2003). Therefore, the challenge relies on creating a context where students can understand LM principles and practices, experiencing relevant technical
and underlying social concepts such as different types of waste, pull production,
balancing, batch reduction, cycle time, work-in-progress, teamwork, communication, etc. (Mazur et al. 2012).

McManus et al. (2007) affirm that teaching LM has proved to be extremely
challenging, especially for engineering students who are not used to abstractions
such as enterprises and process flows. In this sense, some of the leading universities, such as MIT and Virginia Tech, have developed and improved their courses

in the recent years in order to support the teaching/learning activities concerning LM (Badurdeen et al. 2009). Specifically within the industrial engineering education, LM has been mainly approached by the operations management, despite its inherent interfaces with other subjects, such as product development, logistics, and ergonomics (Stier 2003; Alves et al. 2014). Despite the advances in teaching LM principles and concepts, the prevailing utilization of traditional teaching methods, such as lecturing and reading assignments, undermines learning and development of students due to LM's practical character (Kahlen et al. 2011). In fact, Conger and Miller (2014) claim that many students usually have little practical experience with LM principles and techniques, which entails the need for an intuitive understanding of complex enterprises, their intrinsic challenges and specific problems, before the LM transformation can make sense to them (Johnson et al. 2003).

Thus, this paper aims at examining the relationship between different LM teaching methods and students learning preferences. The identification of proper matches between teaching methods and learning preferences (Felder and Silverman 1988) provides means to maximize students' performance and understanding with respect to LM practices and principles. In this sense, professors can better plan and adapt their curriculum in order to assertively develop students and better prepare them for current organizational needs and professional skills demands. To achieve that, 76 graduate students from Industrial Engineering, who participated in two LM courses with different teaching methods were assessed according to their learning preferences and performance. Each LM course had a specific teaching approach: (i) classroom lectures and exercises (classified as traditional teaching methods), and (ii) problem-based learning (PBL) in real-world problems from companies undergoing an LM implementation. Because of the exploratory nature of this research, it is developed a set of formal hypotheses in order to obtain a clearer comprehension around the subject and enable a better understating over the boundary conditions that surround the problem.

2 Literature Background and Hypotheses

2.1 Teaching Methods

A current challenge in education is determining how to present course material so that students not only gain knowledge, but also become self-directed learners who develop problem-solving skills that can be applied in their careers (Sheppard et al. 2008). Traditional teaching methods comprise a model of instruction in which basic skills and facts are taught through direct instruction, while knowledge is transferred from the expert to the novice primarily through lecture or print, requiring physical presence of both student and the teacher (Clark 2006). Traditional teaching methods prevail in most fields of science (Sawyer 2013), being typically teacher-centered methods rather than student-oriented applications and techniques (Dimitrios et al. 2013).

A few examples of these methods include: reading texts and problems, formulating questions, attending lectures and monitor discussions, writing and reply brief or extensive questions and objective type questions, solving short or lengthy unstructured problems and cases, oral presentation of topic and reply to short questions from the audience.

At many large colleges and universities, lectures still seem to be the centerpiece of instruction, where students passively absorb preprocessed information and then regurgitate it in response to periodic multiple-choice exams (McCarthy and Anderson 2000). This approach reinforces learning only at the surface (passive) level rather than at the deep (active) level (Dimitrios et al. 2013). Hence, traditional methods encourage students to concentrate on superficial indicators rather than on fundamental underlying principles, thus neglecting deep (active) learning (Sawyer 2013). Specifically for teaching LM, a recent survey indicated that teaching processes contain many different types of errors that detract from students' learning experience and their perceptions of quality and value (Emiliani 2014a). Another survey identified what constitutes quality teaching from current and former students' perspectives (Emiliani 2014b). Combined, these survey results indicate that traditional teaching methods are unsatisfactory and that students view progressive teaching methods as significantly better.

In turn, active learning refers to experiences in which students are thinking about the subject matter as they interact with the instructor and each other (McCarthy and Anderson 2000). Moreover, active learning involves students and helps them to have an in-depth understanding of the course through induction of practice; i.e., the inductive teaching has better results than productive teaching (Griffin and Care 2015). In this sense, active learning strategies refer to a variety of collaborative classroom activities, ranging from long-term simulations to five-minute cooperative problem-solving exercises (Candido et al. 2007; Rultegde 2016). Thus, instead of facilitating the memorization of large quantities of information, these strategies encourage inquiry and interest as students acquire knowledge and skills (Ramsden 2003), positioning a student-centered approach, maximizing their participation and moving them beyond a superficial, fact-based approach to the subject (Abishova et al. 2014). Particularly for LM courses, previous studies (e.g., Stier 2003; Johnson et al. 2003) have highlighted the utilization of these methods as highly beneficial for students learning. They argue that the active involvement of students helps to connect LM concepts they are expected to apply in their own production run and previous knowledge they have about them. In this sense, students need an intuitive understanding of complex enterprises, their intrinsic challenges, and specific problems, before the LM transformation can make sense to them.

Although interpersonal interaction is a key aspect of active learning, instructors often expect students to acquire relevant knowledge neglecting its inherent interactive content (Badurdeen et al. 2009). In fact, McParland et al. (2004) affirm that, despite active learning methods usually provide higher students' examination scores compared to those on the traditional curriculum, this method does not encourage students to use more effectively their learning styles. Therefore, while logistic necessity (large class sizes) dictates that the lecture format will likely continue to be important in the learning process, this only increases the need for balancing passive with active learning wherever and whenever the possibility arises (McCarthy and Anderson 2000; Bicknell-Holmes and Hoffman 2000). Overall, instructors and professors who designate (short and long) learning goals have to evaluate the theoretical knowledge and abilities (skills) of the students to be able to choose the proper methods that promote an assertive learning (Dimitrios et al. 2013). However, since learning is largely determined by students motivation, to improve the process of teaching/learning as well as the understanding of the subject, specific educational methods should be developed taking into account students' preferences (styles) (Tortorella and Cauchick-Miguel 2017).

2.2 Dimensions of Learning Styles

Learning styles are characteristics cognitive, affective, and psychological behaviors that serve as relatively stable indicators of how learners perceive, interact with, and respond to the learning environment. In essence, learning styles are the way one tends to learn best, comprising the preferred method of taking in, organizing, and making sense of information (Felder and Brent 2005; Puji and Ahmad 2016). Since different people process information in different ways, one learning style is neither preferable nor inferior to another (Cuevas 2015). However, a prevailing paradigm states that learning styles may be obtained from learners' experiences in the learning environment instead of being an innate property of an individual (Khatibi and Khormaei 2016; Francis 2016). In this sense, literature evidences indicate that the teaching methods may enhance the student's learning when they match the category of preexisting interaction preferences; alternatively, learning styles may be changed by the experience with the proposed teaching methods (Felder et al. 1998; Katsioloudis and Fantz 2012; Rultegde 2016).

Continued research has been conducted in previous decades with regards to differences in students' learning styles, which are claimed to be a key factor that accounts for the different efficacies of various teaching methods. Hence, the concept of learning styles has been associated with a wide diversity of students' attributes, entailing the development of several models of learning styles (Kaliská 2012). However, there is no general consensus on the attributes of particular learning styles (Khatibi and Khormaei 2016; Zhang et al. 2016). For instance, in Kolb's (1984) experiential learning model students are classified according to their preferences for (a) concrete experience or abstract conceptualization, and (b) active experimentation or reflective observation. The combination of these classifications results four learning styles, namely: (i) diverger (concrete/reflective); (ii) assimilator (abstract/reflective); (iii) converger (abstract/active); and (iv) accommodator (concrete/active). Later, Honey and Mumford (2000) based on Kolb's theory argued that the main characteristics of students can be integrated into four learning styles: active, reflexive, theoretical, and pragmatic. Further, many researchers have extensively studied the relationship between the learning styles and other individual aspects, such as the interaction effect of students' personality (e.g., Felder et al. 2002; Bayram et al. 2008; Puji and Ahmad 2016) and students' underlying culture and age (e.g., Ariani 2013; Hughes 2016).

Another widely acknowledged model is the Felder-Silverman (Felder and Silverman 1988), which, based on the ILS questionnaire (Felder and Soloman 2004), suggests the definition of students' learning styles according to four dimensions: (i) information perception, (ii) information processing, (iii) information input and (iv) information comprehension. For information perception, preferences can be divided into sensing (sights, sounds, physical sensations) and intuitive (memories, thoughts, insights). Sensors tend to be concrete and practical, while intuitors prefer abstractions, such as mathematical models and theories (Felder et al. 2002; Kaliská 2012). The second dimension is categorized according to the way students process the information; i.e., active learners are more likely to understand facts through engagement in physical activities or discussions, while reflective learners are more introspective and prefer to think about them first. Information input is mainly divided into verbal (written or spoken explanations) and visual (pictures, diagrams, demonstrations) learners (Felder and Spurlin 2005). Finally, the fourth dimension comprehends the way students evolve toward understanding. They can be denoted as sequential learners, who tend to better understand in a logical progression of incremental steps; or global learners, who usually think in a more system-oriented manner and prefer to see the "big picture" first.

2.3 Information Processing and LM Teaching Methods

The dimension denoted as "information processing" comprises the way students prefer retaining the information (Felder and Spurlin 2005). This dimension can be classified into two categories: active and reflective learners. Active learners tend to understand information best by doing something active with it, such as discussing, applying or explaining it to others (Puji and Ahmad 2016). In turn, reflective learners prefer to think about the topic quietly first. In general, active learners usually like group work more than reflective learners, who prefer working alone or with one other person whom they know well (Litzinger et al. 2007). According to Felder and Brent (2005), sitting through lectures without getting to do anything physical but take notes is hard for both processing preferences, but particularly hard for active learners.

In terms of LM learning, Murman et al. (2007) pointed that knowledge is mostly acquired from practice, which contrasts to traditional engineering courses that are based upon knowledge from science and mathematics. This emphasizes the need for practical demonstrations that can be used to impart this kind of knowledge with some credibility. In this sense, active learning methods, such as problem-based learning (PBL), might be a valuable approach to learning how to implement and how to practice LM because it accords with lean's emphasis on teams and on a culture of problem-solving, on learning what to pay attention to, on the value of failure, and on the importance of learning in human development (Badurdeen et al. 2009). Thus, to examine the effect of the relationship between different LM teaching methods and information processing on students' performance, we formulate the following hypotheses:

H1a: When active learning methods are applied for teaching LM, active learners are more likely to present a better performance than reflective learners.
H1b: When traditional teaching methods are applied for teaching LM, reflective learners are more likely to present a better performance than active learners.

2.4   Information Perception and LM Teaching Methods

Information perception can be categorized into sensing and intuitive. Sensing learners (sensors) tend to like learning facts, solving problems by well-established methods and usually dislike complications and surprises. In opposition, intuitive learners (intuitors) often prefer discovering possibilities, relationships and innovation, and dislike repetition (Felder et al. 1998, 2002). Further, sensors are more likely than intuitors to resent being tested on material that has not been explicitly covered in class. According to Zywno (2003), sensors tend to be patient with details and good at memorizing facts; intuitors, in turn, may be better at grasping new concepts and are often more comfortable than sensors with abstractions. Overall, sensors tend to be more practical and careful, and do not like courses that have no apparent connection to the real world. Intuitors tend to work faster and to be more innovative, disliking "plug-and-chug" courses that involve a lot of memorization and routine calculations (Francis 2016).

Regarding LM teaching/learning process, Conger and Miller (2013, 2014) state that active learning methods provide students experience with lean principles and practices, and a sense that they are engaging in real problems. Hence, learning becomes a natural outcome of their engagement and motivation to solve these problems. Achanga et al. (2006) emphasize that some LM practices can only be learned by applying them in real work situations, so-called "learning by doing". Thus, it may be hypothesized that the added value of "learning by doing" may lie in the dialogical process of sharing insights, knowledge, and challenges, which gives context to LM content (Leon et al. 2012). In that way, sensors may benefit from active learning methods, while intuitors may prefer learning LM through traditional teaching methods since these provide a predefined and structured path for learning. To verify those relationships we establish the following hypotheses:

H2a: When active learning methods are applied for teaching LM, sensors are more likely to present a better performance than intuitors.
H2b: When traditional teaching methods are applied for teaching LM, intuitors are more likely to present a better performance than sensors.

2.5   Information Input and LM Teaching Methods

Information input is related to the way students usually remember or capture the information. This dimension can be divided into two categories: verbal or visual learners. Visual learners remember best what they see, such as pictures, diagrams, flow charts, timelines, and demonstrations. In turn, verbal learners get more out of words, either written or spoken explanations (Felder and Brent 2005; Litzinger et al. 2007). However, Francis (2016) indicates that everyone may learn more effectively when information is presented both visually and verbally, which emphasizes the need for applying complementary teaching methods.

In most college classes, LM traditional teaching methods usually offer very little visual information, since students mainly listen to lectures and read material written on chalkboards and in textbooks and handouts (Alves et al. 2014). Unfortunately, most people are visual learners, which means that most students do not get nearly as much as they would if more visual presentation were used in class. On the other hand, previous studies that used active learning methods (e.g., Seddon and Caulkin 2007; Badurdeen et al. 2009; Martens et al. 2010; Suárez-Barraza and Rodríguez-González 2015) have provided diversified ways so students capture the required information through games, simulation, group discussions, and activities, etc. In this sense, the application of active learning methods does not necessarily imply that more visual elements would be involved, hence, indicating that there is not a common understanding if this approach meets student preferences. Since our study comprises the application of PBL within the context of companies undergoing an LM implementation, we argue that group discussions, individual interviews, and brainstorming are likely to be adopted during the problem-solving development, which may favor those students that are more verbal. Further, classroom lectures provided during the other LM course are mainly supported by presentation slides and illustrative videos, mitigating the lack of visual information and possibly meeting visual learners' preferences. Therefore, to verify such association, we suggest the following hypotheses:

H3a: When active learning methods are applied for teaching LM, verbal students are more likely to present a better performance than visual students.
H3b: When traditional teaching methods are applied for teaching LM, visual students are more likely to present a better performance than verbal students.

2.6   Information Comprehension and LM Teaching Methods

Felder and Silverman (1988) have suggested that students may comprehend information through two ways: sequentially or globally. Sequential learners usually gain understanding in linear steps followed by logical connections, hence, they may not fully understand the subject but they can nevertheless do something with it. Further, they may know a lot about specific aspects of a subject, but may have trouble relating them to different aspects of the same subject or to different subjects (Katsioloudis and Fantz 2012). In opposition, global learners are likely to learn in large jumps, absorbing material almost randomly without seeing connections, and then suddenly "getting it". In this sense, they may be able to solve complex problems quickly or put things together in novel ways once they have grasped the big picture; however, these students may have difficulty explaining how they did it (Cuevas 2015; Khatibi and Khormaei 2016).

In active learning methods, such as PBL, LM concepts are introduced by a facilitator or coach as needed, rather than front-loaded into lectures according to the history of a subject or the arrangement of the particular textbook (Emiliani 2015). In this approach, tools and information may be provided by the faculty to solve the problem, but it is the responsibility of students to "make sense" of them by drawing conclusions based on his/her own experience and knowledge (Johnson et al. 2003). In this sense, students are challenged to exhibit and exercise critical thinking skills, starting from an ill-structured problem, finding its root-cause and suggesting/implementing an assertive solution (Tortorella and Cauchick-Miguel 2017). In turn, LM traditional teaching methods often present a predefined syllabus, aiming to provide students a gradual and increasing knowledge about the main concepts, principles, and practices. However, Emiliani (2013) argue that in general both traditional and newer teaching pedagogies, such as active learning, lack a unifying framework or principles to assure focus on students and guide faculty's improvement efforts and decision-making. Therefore, although a few studies indicate that traditional teaching methods may favor sequential learners while active students engagement methods, such as PBL, better support global learners (Stier 2003), empirical evidence validating such associations is still scarce in literature (McManus et al. 2007). Then, to examine these relationships we formulate the hypotheses, as follows:

H4a: When active learning methods are applied for teaching LM, global learners are more likely to present a better performance than sequential learners.
H4b: When traditional teaching methods are applied for teaching LM, sequential learners are more likely to present a better performance than global learners.

3   Method

The proposed method comprises three steps: (i) study context description, (ii) data collection and (iii) data analysis. These steps are better described in the following sections.

3.1 Study Context Description

The study was carried out in a postgraduate program in Industrial Engineering from a Brazilian public university. The program includes four subject areas: (i) operations management, (ii) ergonomics, (iii) product and process development, and (iv) logistics. This program offers both doctoral and master's degrees, and receives approximately 200 applications per year. Each subject area has its own set of courses which are offered during three quarters of the year, starting in March and ending in December. The available teaching time within each quarter totalizes 48 h, distributed over twelve weeks with a one four-hour meeting per week. Originally, research in LM is mainly developed by the operations management area, and two courses are offered on it.

The first course entitled "LM—practices, principles and concepts" focuses on the level of understanding of principles and concepts of LM, and it is offered in the second quarter of the university's academic calendar (June–August). Its syllabus was defined in order to cover the main principles, practices and concepts of three elements of the Toyota's house: basic stability, just-in-time, and *jidoka* (Liker and Meier 2006; Womack and Jones 2009). The teaching methods applied in this course comprise: (i) classroom lectures, (ii) inquiry learning, (iii) class exercises, and (iv) visits and plant tours in manufacturing facilities. Regarding students' performance evaluation, besides classroom participation, which is assessed after each class based on exercises development and corresponds to 30% of the evaluation, a test is applied in the end of the quarter contributing to the remaining 70% of students' final grade, whose scale continuously ranges from 0 to 10.

The second course entitled "LM—PBL in companies undergoing lean implementation" applies PBL approach within a large auto parts manufacturer that is undergoing a lean implementation. Its purpose is to link the concepts and principles to conditions and procedures for application of LM, and it is offered from September to December. This company has been implementing LM for 10 years and has been a partner of the course for three years, opening its site to students and providing the required structure for their development. Students are divided into teams with four or five members and assigned different ill-structured problems previously agreed between the lecturer and company's management. From the twelve weeks scheduled for the course, one meeting aimed at introducing its contents and method, ten meetings are held inside the company for problem-solving activities and data collection, and one final meeting for teams' presentation through an A3 report that follows the Plan-Do-Check-Act (PDCA) process. Final grades are a composition of the assessment of the university lecturer (60% of the final grade) and the company supervisor (40% of the final grade), based on five different competences' criteria (Bédard et al. 2012): (i) knowledge, (ii) analysis, (iii) application, (iv) comprehension and (v) synthesis. The scale for the final grades varied from 0 to 10.

3.2 Data Collection

From 2015 to 2017, 76 graduates students (51 Master candidates and 15 Ph.D. candidates), who started their research in LM, were invited to participate in both courses. From these, 30 students affirmed to have previous experience with LM (either practical or academic), and one-third of the students were full-time dedicated to their research activities (did not work outside the university during their Master or Ph.D.). Thus, the criterion for selecting the sample of students was to include only students who have participated in both LM courses, in this case offered at the same Brazilian public university in the south of Brazil, as to reduce the effects of the external environment (e.g., differences in postgraduate curricula, and emphasis on doctoral and masters research, etc.), since this would be relatively homogeneous within the sample. The nonrandom choice of respondents for surveys and the search for respondents that are already known to the researchers is a commonly used strategy in other studies on LM (e.g., Boyle et al. 2011; Tortorella et al. 2015). For example, Shah and Ward (2007) used a sample with participants drawn from courses and training events when they conducted a survey on LM, since it was necessary that the respondents had experience in the subject.

Further, to collect information about students' characteristics we applied a first questionnaire, which aimed at identifying the demographic profile of the students' sample according to the existence of experience on LM, level of dedication to the postgraduate program (full- or part-time) and level of application (Master of Ph.D.).

Regarding the learning styles, we adapted and applied the Index of Learning Style (ILS) questionnaire with the graduate students in order to verify the preferred learning style (Felder and Soloman 2004; Felder and Spurlin 2005). The ILS is an instrument consisted by 44 questions used to assess preferences on four dimensions of a learning style: (i) information processing (active/reflective), (ii) information perception (sensing/intuitive), (iii) information input (visual/verbal), and (iv) information comprehension (sequential/global). The questionnaire results simply indicate preferences for each dimension, and the suggestions that follow the results may enable to verify the match with the teaching method (Felder et al. 1998; Litzinger et al. 2007). For each learning dimension, a cardinal scale ranging from 1 to 5 was proposed (see Fig. 1), varying between the emphasis on the respective dimension categories, such as: strong preference (1 or 5), moderate preference (2 or 4) and fairly well-balanced (3). The evaluation of the learning styles was carried out in both courses to check for inconsistencies. We did not find any inconsistency or changes in the learning style of any of the students analyzed.

Finally, students' performance during both courses was obtained from the respective final grades. However, since the evaluation methods employed in each course was different, to better compare students' performance between them we standardized those grades in terms of the number of standard deviations of each individual value with respect to the average grade of the respective course. The standardized grades enable to remove scale effects (Tortorella and Fogliatto 2014), and large positive values indicate the best performances of each course.

Learning dimension	Strong preference	Moderate preference	Fairly well-balanced	Moderate preference	Strong preference
	1	2	3	4	5
Information processing	ACTIVE ◄┈┈┈┈┈┈┈┈┈┈┈┈┈┈┈┈┈┈┈┈┈┈┈┈┈┈┈► REFLECTIVE				
Information perception	SENSING ◄┈┈┈┈┈┈┈┈┈┈┈┈┈┈┈┈┈┈┈┈┈┈┈┈┈► INTUITIVE				
Information input	VISUAL ◄┈┈┈┈┈┈┈┈┈┈┈┈┈┈┈┈┈┈┈┈┈┈┈┈┈► VERBAL				
Information comprehension	SEQUENTIAL ◄┈┈┈┈┈┈┈┈┈┈┈┈┈┈┈┈┈┈┈┈► GLOBAL				

Fig. 1 Learning style dimensions and assessment scales. *Source* Adapted from Felder et al. (1998)

Table 1 Sperman's correlations between learning dimensions and students' experience on LM*

Variables	Previous experience	Information processing	Information perception	Information input	Information comprehension
Previous experience	–	0.06	0.08	0.70	0.84
Information processing	−0.22	–	0.16	0.42	0.06
Information perception	0.20	−0.16	–	0.61	0.06
Information input	0.05	0.01	0.06	–	0.99
Information comprehension	−0.02	−0.22	0.24	0.00	–

*Values below diagonal are Spearman's correlations and values above diagonal are p-values

A correlation matrix for each of the learning dimensions and students' experience on LM included in the analysis is shown in Table 1. Based on this analysis, no significant correlation (p-value $\leq$ 0.05) was identified, which enabled the analysis of each independent variable disregarding multicollinearity effects.

3.3 Data Analysis

The data analysis was performed using the SPSS® version 24 statistical software. The 76 answers were tested according to the effect of the dimensions of learning styles (independent variables) on students' performance (dependent variable) for each teaching method. Previous experience, which was coded into existing (1) or not (0), was used as a control variable for our study's purpose. To achieve that, a stepwise regression with backward elimination was performed. Stepwise-backward regression is a method of fitting regression models in which the choice of predictive variables is carried out by an automatic procedure (Hair et al. 2006). In each step, a variable is considered for subtraction from the set of explanatory variables based on some prespecified criterion. Usually, this takes the form of a sequence of F-tests or t-tests, but other techniques are possible, such as adjusted R^2. This method is usually recommended for exploratory studies in which it is intended to describe little-known relationships between the variables (Harrell 2001).

Therefore, we started involving all candidate variables (independent and control variables) to predict students' performance for each teaching method, testing the deletion of each one using the F-test criterion. Whenever the statistical significance (p-value ≤ 0.05) of the F-test was not achieved, we deleted the variable (if any) whose loss gives the most statistically insignificant deterioration of the model fit, and repeated this process until no further variables can be deleted without a statistically significant loss of fit.

Assumptions of normality, linearity, and homoscedasticity were tested for the learning dimensions. We examined the residuals to confirm normality of the error term distribution. Linearity was tested with plots of partial regression for each dimension. Finally, we evaluated homoscedasticity by plotting standardized residuals against predicted value and examining visually. These tests confirmed the three assumptions for multivariate data analysis.

4   Results

Regarding students' performance with the PBL method for teaching LM, Table 2 presents the ANOVA for the regression analysis of the dimensions of learning styles and the PBL method (active learning). The solution evolved through four models, following the backward elimination logic. Model 4a achieved the required statistical significance (p-value ≤ 0.05), explaining 36% of total variance. This model indicates that the variables "information processing" and "information perception" are capable to predict students' performance when active learning methods, such as PBL, are used to teach LM. Further, coefficients of all regression models are shown in Table 3, along with the collinearity statistics. We also calculated VIF (variance inflation factors) for all coefficients in the regression models. Previous researchers (e.g., Hair et al. 2006; Hayes and Matthes 2009; Tabachnick and Fidell 2013) suggest that values below five indicate that multicollinearity problems do not affect the coefficient. The VIF for all coefficients of Model 4a attended such requirement, allowing the regression model to attribute performance effects to all individual variables. Thus, from the four hypotheses related to the application of active learning methods to teach LM (e.g., PBL), we found evidence to support *H1a* and *H2a*, while *H3a* and *H4a* could not be verified with our study sample.

Table 2 ANOVA results for PBL method

Model	Predictors		Sum of squares	df	Mean square	F	p-value
1a	Previous experience Information processing Information perception Information input Information comprehension	Regression	7.674	5	1.535	1.636	0.162
		Residual	65.660	70	0.938		
		Total	73.334	75			
2a	Previous experience Information processing Information perception Information input	Regression	7.666	4	1.916	2.072	0.094
		Residual	65.668	71	0.925		
		Total	73.334	75			
3a	Previous experience Information processing Information perception	Regression	7.593	3	2.531	2.772	0.058
		Residual	65.741	72	0.913		
		Total	73.334	75			
4a	Information processing Information perception	Regression	6.371	2	3.185	3.473	0.036
		Residual	66.963	73	0.917		
		Total	73.334	75			

Information processing appears to be negatively associated with students' performance ($\beta = -0.273$), which suggests that students who prefer to retain the information by discussing or applying it (active learners) are more likely to perform better under a PBL approach. Since the second course (LM—PBL in companies undergoing lean implementation) provided students the opportunity to interact with real-world problems to which they were pushed to indicate a practical and effective solution through a PDCA-thinking process, it is quite reasonable that active learners may benefit from this teaching approach and present a higher performance. Therefore, this result supports *H1a* and converges to findings from previous studies on LM (e.g., Candido et al. 2007; Aij et al. 2013; Emiliani 2015) that have emphasized the necessity of "learning by doing", but did not verified empirically such indication.

With regards to information perception, our results showed that this dimension presents a significant negative association ($\beta = -0.256$) with students' performance when PBL method is employed to teach LM. In other words, this outcome indicates

Table 3 Regression models for PBL method

Model	Predictors	β	Std. error	p-value	Tolerance	VIF	R^2	Adj. R^2
1a	(Constant)	−1.430	0.742	0.058			0.403	0.395
	Previous experience	−0.268	0.236	0.260	0.931	1.074		
	Information processing	−0.241	0.141	0.093	0.896	1.116		
	Information perception	−0.268	0.136	0.052	0.870	1.149		
	Information input	0.036	0.125	0.777	0.984	1.016		
	Information comprehension	0.014	0.154	0.928	0.874	1.145		
2a	(Constant)	−1.391	0.594	0.022			0.398	0.387
	Previous experience	−0.269	0.233	0.252	0.936	1.068		
	Information processing	−0.239	0.138	0.089	0.925	1.081		
	Information perception	−0.272	0.129	0.039	0.948	1.055		
	Information input	0.035	0.123	0.779	0.988	1.012		
3a	(Constant)	−1.325	0.543	0.017			0.388	0.376
	Previous experience	−0.268	0.232	0.251	0.936	1.068		
	Information processing	−0.242	0.137	0.081	0.932	1.073		
	Information perception	−0.275	0.128	0.035	0.954	1.048		
4a	(Constant)	−1.463	0.532	0.007			0.375	0.360
	Information processing	−0.273	0.134	0.046	0.970	1.031		
	Information perception	−0.256	0.127	0.048	0.970	1.031		

that students who are sensors (either strong or moderate preference) are quite likely to learn LM more effectively when they have a chance to connect their knowledge on LM principles and practices to real-world situations. This result reinforces the LM principle of *genchi-genbutsu* (go and see), which states that the inference of any information is legitimized when the individual has the chance to see with their own eyes, establishing empathy with the analyzed context (Liker and Meier 2006; Womack and Jones 2009). Thus, our findings bear *H2a*.

For traditional teaching methods of LM, Tables 4 and 5 display the results of the ANOVA for the regression analysis of the dimensions of learning styles and their variables' coefficients, respectively. Similarly, the solution evolved through four

Table 4 ANOVA results for traditional teaching method

Model	Predictors		Sum of squares	df	Mean square	F	p-value
1b	Previous experience Information processing Information perception Information input Information comprehension	Regression	7.031	5	1.406	1.780	0.128
		Residual	55.284	70	0.790		
		Total	62.315	75			
2b	Information processing Information perception Information input Information comprehension	Regression	7.024	4	1.756	2.255	0.072
		Residual	55.291	71	0.779		
		Total	62.315	75			
3b	Information perception Information input Information comprehension	Regression	6.825	3	2.275	2.952	0.058
		Residual	55.490	72	0.771		
		Total	62.315	75			
4b	Information perception Information comprehension	Regression	5.244	2	2.622	3.353	0.040
		Residual	57.071	73	0.782		
		Total	62.315	75			

models in a backward elimination, in which Model 4b attended the required significance level (p-value $\leq$ 0.05), explaining 32.1% of total variation. This model suggests that "information perception" and "information comprehension" can predict students' performance when traditional methods, such as classroom lectures, are applied to teach LM. Further, VIF requirements were achieved, as aforementioned, indicating no multicollinearity problems effect on the coefficients of Model 4b. Therefore, although hypotheses *H1b* and *H3b* were not supported by this outcome, we found evidence to bear *H2b* and *H4b* with our study sample.

Specifically for information perception, contrary to what was found to PBL method, this dimension seems to be positively associated (β = 0.249) with students' performance undergoing an LM course based on traditional teaching methods. This means that intuitive learners, who are usually comfortable with abstractions and dislike repetition, may learn more effectively LM principles and concepts when methods such as classroom lectures and exercises, inquiry learning and shop floor visits are employed. Further, according to Felder and Silverman (1988), intuitors prefer principles and theories which were the main focus of the course "LM—practices, principles and concepts", favoring such performance trend with traditional teaching methods. In this sense, our result corroborates with the studies from

Table 5 Regression models for traditional teaching method

Model	Predictors	β	Std. error	p-value	Tolerance	VIF	R^2	Adj. R^2
1b	(Constant)	−0.793	0.681	0.248			0.493	0.451
	Previous experience	0.020	0.216	0.926	0.931	1.074		
	Information processing	0.066	0.130	0.612	0.896	1.116		
	Information perception	0.258	0.125	0.042	0.870	1.149		
	Information input	0.157	0.114	0.174	0.984	1.016		
	Information comprehension	−0.320	0.141	0.026	0.874	1.145		
2b	(Constant)	−0.780	0.662	0.243			0.462	0.404
	Information processing	0.064	0.126	0.615	0.936	1.068		
	Information perception	0.257	0.123	0.040	0.888	1.126		
	Information input	0.157	0.114	0.170	0.984	1.016		
	Information comprehension	−0.319	0.140	0.025	0.878	1.138		
3b	(Constant)	−0.570	0.514	0.271			0.378	0.350
	Information perception	0.264	0.121	0.032	0.902	1.109		
	Information input	0.161	0.113	0.156	0.989	1.011		
	Information comprehension	−0.307	0.137	0.028	0.903	1.107		
4b	(Constant)	−0.185	0.441	0.676			0.344	0.321
	Information perception	0.249	0.121	0.044	0.909	1.100		
	Information comprehension	−0.292	0.137	0.037	0.909	1.100		

Mazur et al. (2012) and Suárez-Barraza and Rodríguez-González (2015), which reinforced that learning LM requires training in both soft and hard skills in order to solve problems in both the sociocultural and technical aspects of a value stream. Hence, traditional teaching methods may support such learning by providing students minimum ground knowledge that entails changes in their understanding about LM. Thus, our results empirically validate *H2b*.

Results for information comprehension indicate that this learning dimension has a significant negative association ($\beta = -0.292$) with traditional teaching methods. In fact, while sequential learners may benefit from these teaching methods, global learners might find difficult to understand LM concepts and principles under this approach, hence, presenting a worse performance in this course. This result is somewhat consistent with findings from Aij et al. (2013) and Tortorella and Cauchick-Miguel (2017), which suggest that students' lack of experience in the LM principles and practices usually features a barrier to transferring the knowledge from a usual LM course into practical situations. In this sense, the expectation that low-experienced students may learn in large jumps, absorbing material almost randomly without seeing connections, and then suddenly "getting it", becomes contradictory. Thus, our result supports *H4b* and corroborates with previous studies' indications.

Overall, from the eight hypotheses formulated for this study our results verified four of them, being two for each LM teaching method applied (see Fig. 2). Further, the effect of the learning dimension "information perception" seems to be more extensive than the others, since students' performance in both teaching methods is significantly associated with it. Thus, lecturers and instructors might consider students' perception preferences when planning their LM teaching approach. Further, this outcome can be related to one of the main paradoxal characteristics of LM. According to Spear and Bowen (1999) and Spear (2009), the high level of process standardization is exactly what encourages flexibility and innovation in a lean system. In terms of teaching efficiency, this concept can be extended to sensors and intuitors preferences. If instructors and lecturers overemphasize intuition and reinforce only high-innovative activities, students may miss important details or make careless mistakes in calculations or hands-on work; in turn, if they overemphasize sensing and provide only standardized and widely acknowledged methods, the other part of the students may rely too much on memorization and familiar methods and not concentrate enough on understanding and innovative thinking. On the other hand, surprisingly, the learning dimension "information input" was not significantly associated with any of the teaching methods. This result may indicate that, despite PBL and traditional teaching methods may offer different ways to students capture the concepts and principles of LM, both methods may interchangeably reinforce visual and verbal mechanisms, entailing a well-balanced approach for this dimension.

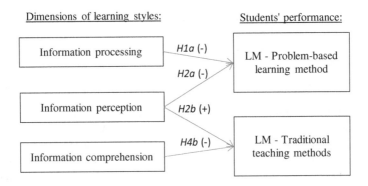

Fig. 2 Hypotheses verified by the models

5 Conclusions

In this paper, we have studied the relationship between LM teaching methods and the dimensions of learning styles in order to maximize students' performance. This research suggests implications for both theory and practice, which are deeper discussed in the following sections.

5.1 *Implications to Theory*

In theoretical terms, our study suggests that mixing active learning methods with traditional teaching methods may facilitate dialogical learning, encourage collaboration between students and facilitate the transfer of knowledge on LM. Current teaching approaches tend to focus on a single teaching/learning point, disregarding the didactic potential of complementary methods. Thus, instructors and lecturers must take into account that blending the student interest generated by LM real-world applications through PBL with some of the structure of traditional teaching methods could increase both student satisfaction and achievement. To determine an effective blend of the two methods it is essential to understand students' learning preferences. Although being impossible to simultaneously tailor the teaching method according to each student learning style in a class, such understanding allows academics to plan the LM course with a higher likelihood of contributing to the students' preferred learning styles. Particularly for LM education, our findings provide evidence to support the application of a hybrid teaching method, in which the traditional teaching methods lay the fundamental concepts followed by a culminating period that provides an opportunity to extend the acquired knowledge into real and practical problems. Therefore, the proper educational approach for teaching and learning of LM would help enhancing students' capability for acquiring and applying knowledge in real-world situations, preparing them to meet the required competencies that fulfill either organizations or academia current demands.

It is noteworthy that, contrary to expectations, improved academic performance was found without any changes in students' learning preferences throughout the courses, and was unrelated to students' previous experience. Consistent with previous research, academic success was related to the use of strategic and deep learning styles. Further, the incorporation of active learning methods, such as PBL, improved students' ability to learn LM, but did not lead to a change in students' preferred approach to learning outside the sessions. In this sense, results indicate that students' learning preferences are a key element for designing LM teaching methods. In fact, these must be systematically followed and widely used by industrial engineering postgraduate programs in order to provide better learning results and pedagogical assertiveness.

5.2 Implications to Practice

With regards to practical implications, there are several benefits for various stakeholders when improving LM course's design/delivery and overall quality of the learning experience. First, an enhanced LM course design and delivery occurs when an interdisciplinary course replicates workplace engagement, allowing the actual employment of LM practices and principles. Second, an improved effectiveness of learning occurs as active learning methods, such as PBL, are incorporated to the curriculum of the course, enabling to extend the pervasiveness of the information among students. In addition, an improved LM course entails benefits to four audiences: (i) students benefit from an enriched learning experience that better prepares them to future organizations' demands; (ii) faculty benefits through the use of enlightened pedagogy that incorporates high-impact teaching methods; (iii) university is affected by short and long-term business/community relationships developed through exchanges of thought leadership and student field experiences; and (iv) employers benefit through assisting students to align with employee development expectations, based on workplace gaps, such as learning to communicate and exercise leadership.

Further, traditional teaching methods when coupled with implementation through real-world situations, not only can students better understand the abstract and sometimes counter-intuitive LM concepts and how to implement them, but they are given the opportunity to develop a broad range of soft-skills. In other words, because it foregrounds the social context for problem-solving and learning, LM learning must be experiential. Thus, a proper LM education not only teaches professional skills necessary for industry, but creates a platform for transition between university and the working world. This bridging-the-gap between institution and industry has undoubtedly aided in the student's preparation as an engineer.

5.3 Limitations and Future Research

The LM practices, principles, and concepts can be extended and correlated to processes and systems within many engineering disciplines and are not limited to manufacturing or operations management. Regardless of the discipline, traditional and newer teaching pedagogies for teaching LM can have the appearance of effectiveness based on empirical evidence such as test scores or students feedback. However, these pedagogies may continue using assessment methods that limit or reduce students learning. Therefore, future studies that include students' evaluation methods may provide a more robust comparison of students' performance in different LM courses, avoiding any misguided parameter or assessment for such inference.

Additionally, future challenges facing both students and LM education demand fundamental yet practical reforms. Instructors and lecturers may continue to reject improvement methods born in industry, or they can begin the process of scholarly inquiry, for which they are well equipped, to understand LM principles and practices and how to apply them to improve teaching itself. Moreover, regarding the relationship between the ILS and the LM teaching methods, future research is needed to expand the investigation to other teaching methods besides PBL and traditional methods. Thus, a better comprehension of the concurrent influence of several teaching methods and students' learning preferences may provide arguments to professors and academics to anticipate their synergistic effect, so the end results are more likely to meet the expectations.

References

Abishova, G., Bostanova, A., Isaev, A., Erimova, A., Salybekova, N., Serzhanova, A., et al. (2014). Teaching practice using interactive methods at the higher educational establishments. *Procedia—Social and Behavioral Sciences, 143,* 630–633.

Achanga, P., Shehab, E., Roy, R., & Nelder, G. (2006). Critical success factors for lean implementation within SMEs. *Journal of Manufacturing Technology Management, 17*(4), 460–471.

Aij, K., Simons, F., Widdershoven, G., & Visse, M. (2013). Experiences of leaders in the implementation of Lean in a teaching hospital-barriers and facilitators in clinical practices: A qualitative study. *British Medical Journal Open, 3,* e003605.

Alves, A., Kahlen, F., Flumerfelt, S., & Manalang, A. (2014). The lean production multidisciplinary: From operations to education. In *Proceedings of 7th International Conference on Production Research, Americas.*

Ariani, D. (2013). Personality and learning motivation. *European Journal of Business and Management, 5,* 10–26.

Badurdeen, F., Marksberry, P., Hall, A., & Gregory, B. (2009). Teaching lean manufacturing with simulations and games: A survey and future directions. *Simulation & Gaming, 4*(4), 465–486.

Bayram, S., Deniz, L., & Erdoğan, Y. (2008). The role of personality traits in web-based education. *The Turkish Online Journal of Educational Technology–TOJET, 7*(2), 5–41.

Bédard, D., Lison, C., Dalle, D., Côté, D., & Boutin, N. (2012). Problem-based and project-based learning in engineering and medicine: Determinants of students' engagement and persistence. *Interdisciplinary Journal of Problem-Based Learning, 6*(2), 7–30.

Bicknell-Holmes, T., & Hoffman, P. (2000). Elicit, engage, experience, explore: Discovery learning in library instruction. *Reference Service Review, 28*(4), 313–322.

Boyle, T., Scherrer-Rathje, M., & Stuart, I. (2011). Learning to be lean: The influence of external information sources in lean improvements. *Journal of Manufacturing Technology Management, 22*(5), 587–603.

Candido, J., Murman, E., & McManus, H. (2007). Active learning strategies for teaching lean thinking. In *Proceedings of the 3rd International CDIO Conference,* Cambridge, MA, June 11–14.

Clark, A. (2006). Changing classroom practice to include the project approach. *Early Childhood Research & Practice, 8*(2).

Conger, S., & Miller, R. (2013). *Problem-based learning for a lean six sigma course,* University of Dallas, USA. Sprouts: Working Papers on Information Systems, 13(1).

Conger, S., & Miller, R. (2014). Problem-based learning applied to student consulting in a lean production course. *Journal of Higher Education Theory and Practice, 14*(1), 81.

Cuevas, J. (2015). Is learning styles-based instruction effective? A comprehensive analysis of recent research on learning styles. *Theory and Research in Education, 13*(3), 308–333.

Dimitrios, B., Labors, S., Nikolaos, K., Maria, K., & Athanasios, K. (2013). Traditional teaching methods vs. teaching through the application of information and communication technologies in the accounting field: Quo vadis? *European Scientific Journal, 9*(28), 73–101.

Dukovska-Popovska, I., Hove-Madsen, V., & Nielsen, K. (2008). Teaching lean thinking through game: Some challenges. In *Proceedings of the 36th European Society for Engineering Education (SEFI) on Quality Assessment, Employability & Innovation.*

Emiliani, B. (2013). *The lean professor: Become a better teacher using lean principles and practices.* LLC, Wethersfield, CT: The CLBM.

Emiliani, B. (2014a). Teaching survey's—interim results. *The Lean Professor blog post,* 5 January. Available at: http://leanprofessor.com/blog/2014/01/05/teaching-surveys-interim-results/. Accessed on June 20, 2017.

Emiliani, B. (2014b). What is good quality teaching?—Survey results. *The Lean Professor blog post,* 13 February. Available at: http://leanprofessor.com/blog/2014/02/13/good-quality-teaching-survey-results/. Accessed on June 20, 2017.

Emiliani, M. (2015). Engaging faculty in lean teaching. *International Journal of Lean Six Sigma, 6*(1).

Felder, R., & Brent, R. (2005). Understanding student differences. *Journal of Engineering Education, 94*(1), 57–72.

Felder, R., Felder, G., & Diez, E. (1998). A longitudinal study of engineering student performance and retention versus comparisons with traditionally-taught students. *Journal of Engineering Education, 87*(4), 469–480.

Felder, R., Felder, G., & Dietz, E. (2002). The effects of personality type on engineering student performance and attitudes. *Journal of Engineering Education, 91*(1), 3–17.

Felder, R., & Silverman, L. (1988). Learning and teaching styles in engineering education. *Engineering Education, 78*(7), 674–681.

Felder, R., & Soloman, B. (2004). *Index of Learning Styles (ILS).* Available at: www2.ncsu.edu/unity/lockers/users/f/felder/public/ILSpage.html. Accessed on May 5, 2016.

Felder, R., & Spurlin, J. (2005). Applications, reliability and validity of the index of learning styles. *International Journal of Engineering Education, 21*(1), 103–112.

Francis, R. (2016). Learning styles: Key to enhance learning among student teachers of the B. ED course. *International Education and Research Journal, 2*(12), 54–55.

Griffin, P., & Care, E. (2015). *Assessment and teaching of 21st century skills: Methods and approach.* Melbourne, Australia: Springer.

Hair, J., Tatham, R., Anderson, R., & Black, W. (2006). *Multivariate data analysis.* NJ, Pearson Prentice Hall: Upper Saddle River.

Harrell, F. (2001). *Regression modeling strategies: With applications to linear models, logistic regression, and survival analysis.* New York: Springer-Verlag.

Hayes, A., & Matthes, J. (2009). Computational procedures for probing interactions in OLS and logistic regression: SPSS and SAS implementations. *Behavior Research Methods, 41*(3), 924–936.

Honey, P., & Mumford, A. (2000). *The learning styles helper's guide.* First published on September 2000, revised edition April 2006. Available at: www.peterhoney.com.

Hughes, G. (2016). Identifying learning styles: A CPD article improved Grace Hughes's knowledge of how to identify different learning styles. *Nursing Standard, 31*(16–18), 72–73.

Johnson, S., Gerstenfeld, A., Zeng, A., Ramos, B., & Mishra, S. (2003). Teaching lean process design using a discovery approach. In *Proceedings of the American Society for Engineering Education Annual Conference and Exposition.*

Kahlen, F., Flumerfelt, S., Siriban-Manalang, A., & Alves, A. (2011). Benefits of lean teaching. In *Proceedings of American Society of Mechanical Engineers (ASME) International Mechanical Engineering Congress & Exposition* (pp. 12–18).

Kaliská, L. (2012). Felder's learning style concept and its index of learning style questionnaire in the Slovak conditions. *Grant Journal, 1,* 52–56.

Kanigolla, D., Cudney, E., & Corns, S. (2014). Enhancing engineering education using project-based learning for lean and six sigma. *International Journal of Lean Six Sigma, 5*(1), 45–61.

Karlsson, C., & Ahlstom, P. (1996). Assessing changes towards lean production. *International Journal of Operations & Production Management, 16*(2), 24–41.

Katsioloudis, P., & Fantz, T. (2012). A comparative analysis of preferred learning and teaching styles for engineering, industrial, and technology education: Students and faculty. *Journal of Technology Education, 23*(2), 61–69.

Khatibi, M., & Khormaei, F. (2016). Learning and personality: A review. *Journal of Educational and Management Studies, 6*(4), 89–97.

Kolb, D. (1984). *Experiential learning: Experience as the source of learning and development.* Englewood Cliffs, NJ: Prentice-Hall.

Leon, H., Perez, M., Farris, J., & Beruvides, M. (2012). Integrating six sigma tools using team-learning processes. *International Journal of Lean Six Sigma, 3*(2), 133–156.

Liker, J., & Meier, D. (2006). *The Toyota way field book: A practical guide for implementing Toyota's 4Ps.* New York: McGraw-Hill.

Litzinger, T., Lee, S., Wise, J., & Felder, R. (2007). A psychometric study of the index of learning styles[©]. *Journal of Engineering Education, 96*(4), 309.

Martens, I., Colpaert, J., & De Boeck, L. (2010). Lean learning academy: An innovative learning concept in engineering curricula. In *Proceedings of IHEPI 2010 Conference Paper*, Budapest, Hungary.

Mazur, L., McCreery, J., & Rothenberg, L. (2012). Facilitating lean learning and behaviors in hospitals during the early stages of lean implementation. *Engineering Management Journal, 24* (1), 11–22.

McCarthy, J., & Anderson, L. (2000). Active learning techniques versus traditional teaching styles: Two experiments from history and political science. *Innovative Higher Education, 24* (4), 279–294.

McManus, H., Rebentisch, E., Murman, E., & Stanke, A. (2007). Teaching lean thinking principles through hands-on simulations. In *Proceedings of the 3rd International CDIO Conference*, MIT, Cambridge, Massachusetts, June 11–14.

McParland, M., Noble, L., & Livingston, G. (2004). The effectiveness of problem-based learning compared to traditional teaching in undergraduate psychiatry. *Medical Education, 38,* 859–867.

Murman, E., McManus, H., & Candido, J. (2007). Enhancing faculty competency in lean thinking bodies of knowledge. In *Proceedings of the 3rd International CDIO Conference*, Cambridge, MA, June 11–14.

Puji, R., & Ahmad, A. (2016). Learning style of MBTI personality types in history learning at higher education. *Scientific Journal of PPI-UKM, 3*(6), 289–295.

Ramsden, P. (2003). *Learning to teach at higher education.* London: Taylor and Francis.

Rultegde, S. (2016). *What do teachers know about differentiated instruction?* (Doctoral dissertation, Graduate Faculty of the Louisiana State University and Agricultural and Mechanical College, Southern University).

Sawhney, R., & Chason, S. (2005). Human behavior based exploratory model for successful implementation of lean enterprise in industry. *Performance Improvement Quarterly, 18*(2), 76–96.

Sawyer, G. (2013). *The effects of traditional teaching methods, project-based learning, and a blended teaching style on elementary students* (Ph.D. Dissertation, Faculty of Trevecca Nazarene, University School of Education).

Seddon, J., & Caulkin, S. (2007). Systems thinking, lean production and action learning. *Action Learning: Research and Practice, 4*(1), 9–24.

Shah, R., & Ward, P. (2007). Defining and developing measures of lean production. *Journal of Operations Management, 25*(4), 785–805.

Sheppard, S., Macatangay, K., Colby, A., Sullivan, W., & Shulman, L. (2008). *Educating engineers: Designing for the future of the field.* Hoboken, NJ: Jossey-Bass.

Spear, S. (2009). *The high-velocity edge: How market leaders leverage operational excellence to beat the competition*. New York, NY: McGraw-Hill.

Spear, S., & Bowen, H. (1999). Decoding the DNA of the Toyota production system. *Harvard Business Review, 77,* 96–108.

Stier, K. (2003). Teaching lean manufacturing concepts through project-based learning and simulation. *Journal of Industrial Technology, 19*(4), 1–6.

Suárez-Barraza, M., & Rodríguez-González, F. (2015). Bringing Kaizen to the classroom: Lessons learned in an operations management course. *Total Quality Management & Business Excellence, 26*(9–10), 1002–1016.

Tabachnick, B., & Fidell, L. (2013). *Using multivariate statistics* (5th ed.). New York, NY: Pearson.

Tortorella, G., & Cauchick-Miguel, P. (2017). An initiative for integrating problem-based learning into a lean manufacturing course of an industrial engineering graduate program. *Production, 27* (Special Issue), e20162247.

Tortorella, G., & Fogliatto, F. (2014). Method for assessing human resources management practices and organisational learning factors in a company under lean manufacturing implementation. *International Journal of Production Research, 52*(15), 4623–4645.

Tortorella, G., Maroding, G., Miorando, R., & Fogliatto, F. (2015). Learning organisation and human resources management practices: An exploratory research in medium-sized enterprises undergoing a lean implementation. *International Journal of Production Research, 53*(13), 3989–4000.

Wan, H., & Chen, F. (2008). A leanness measure of manufacturing systems for quantifying impacts of lean initiatives. *International Journal of Production Research, 46*(23), 6567–6584.

Womack, J., & Jones, D. (2009). *Lean solutions: How companies and customers can create value and wealth together*. New York: Simon and Schuster.

Zhang, J., Ogan, A., Liu, T., Sung, Y., & Chang, K. (2016). The influence of using augmented reality on textbook support for learners of different learning styles. In *Proceedings of the 2016 IEEE International Symposium on Mixed and Augmented Reality (ISMAR), IEEE* (pp. 107–114).

Zywno, M. (2003). A contribution to validation of score meaning for Felder-Soloman's index of learning styles. In *Proceedings of the American Society for Engineering Education Annual Conference & Exposition*.